THE
BEATING
HEART

THE BEATING HEART

The Art & Science of Our Most Vital Organ

Robin Choudhury

HEAD
ZEUS

An Apollo Book

First published in the UK in 2024 by Head of Zeus,
part of Bloomsbury Publishing Plc

9 7 5 3 1 2 4 6 8

A catalogue record for this book is available from the British Library.

ISBN (HB): 9781837931767
ISBN (E): 9781837931743

Designed and typeset by Heather Bowen
Colour separation by Dawkins Colour
Printed and bound in Turkey by Elma Basim

Head of Zeus Ltd
First Floor East
5–8 Hardwick Street
London EC1R 4RG

www.headofzeus.com

Cordis et Sanguinis
(*Of the heart and of the blood,* after William Harvey)

For Jasmine
and for Bert, Maya, Sophia and Milo

Contents

I

The Ancient and Pervasive Heart

*The heart is situated in the middle of the body
and in the fore part of it; and in the heart,
we hold, is the principle of life and of all
movement and sensation.*

ARISTOTLE
De Partibus Animalium (*Parts of Animals*), *c.* 350 BCE

Where to search for the first stirrings of life?

Aristotle looked inside an egg.

Fashioning a window in the shell of a chick egg a few days after fertilization, he revealed something remarkable. Visible on the surface, even before the larger anatomical structures had formed, he saw the embryonic heart, red with life blood, beating rhythmically, regularly, at the hub of a network of blood vessels. The self-evident originator and provider of life.

Strip away modern knowledge of biochemistry, genes, cells and microscopes, embryology and reproductive biology and simply imagine, for a moment, the profundity of believing that you have identified the very origin of life. Little wonder that Aristotle's conceptual thinking in biology, and beyond, was so heavily built around his ideas about the heart. Importantly, inevitably, unlike his predecessors, Aristotle placed the heart at the *centre* of the body.

Natural philosopher, systematic observer and proto-scientist, Aristotle (*c.* 384 – *c.* 322 BCE) had started his scholastic life in Athens, under Plato. Both of his parents were members of traditional medical families, and his father had served as court physician to the king of Macedonia. At the age of seventeen, Aristotle was sent to Athens, to Plato's Academy. He spent twenty years as a student and teacher at the school, but around 348 BCE he left Athens abruptly, possibly as a consequence of an academic dispute or other local political strife, and eventually came to settle on the Aegean island of Lesbos, off the coast of modern Turkey.[1] There, he began an unprecedented systematic examination of living creatures. He committed his observations and syntheses to written texts, but without illustration. He was among the first to record the dissection of animals, and he certainly knew that after death the heart became limp and still. He would, of course, have experienced the beating of his own heart, and its quickening in response to physical exertion, passion and intense emotion. How could he fail to notice that in life most creatures are warm, with beating hearts, and that in death the heart stops and the warmth is lost? How could he make sense of the knowledge that breath is heated during its passage into and out of our lungs? How could he possibly explain that the core of the body is hot, while the peripheries are cooler? Drawing together these fundamental observations, how could Aristotle *not* conclude that the heart was the source of life?

In *De anima* (*Of the Soul*), he reasoned that anything that nourishes itself, anything that grows, anything that is capable of locomotion, perception or thought, is alive – and that the driver, or provider, and integrator of these capabilities is the 'soul'. The soul was causally responsible for animate behaviour and, for Aristotle, it resided in the heart; indeed, it was inseparable from the heart. All these characteristics were integral properties of the organ itself.

In *De generatione animalium* (*Generation of Animals*), he wrote: 'the first principle of any natural creature's system is the heart or its counterpart', and in *De partibus animalium* (*Parts of Animals*), he wrote: 'The heart is situated in the middle of the body and in the fore part of it; and in the heart, we hold, is the principle of life and of all movement and sensation.'

The idea that the soul was located in the heart (as Aristotle thought), rather than in the brain (Plato's doctrine), was long contested. As we shall see, equivalent arguments resurfaced in the seventeenth-century tension between William Harvey and René Descartes regarding the pre-eminence of the heart over the brain.

Our contemporary notions of the 'soul' have evolved and, influenced by later schools of philosophical and theological thought, have attracted various connotations that would not have been recognized by Aristotle. His 'soul' was not the later, morally responsible individual soul of Thomas Aquinas. It was not immortal; it did not survive outside the body. What emerged from Aristotle was a complex amalgamation, wherein the heart was at once the seat of the soul, the originator of life, the hub of perception and the source of all bodily function. By extension, all living organisms had souls, the complexity of which, he believed, accorded with the sophistication of the organism.

Aristotle's ideas on the centrality of the heart were buried for hundreds of years but resurfaced in the twelfth century through the work of Arab scholars. His ideas played a prominent, though not unchallenged, role in Western thinking about the function and significance of the heart, up to and including the work of William Harvey – a span of 2,000 years.

Indeed, part of my motivation for researching this book has been to examine how the heart has been regarded through time and across cultures, drawing on an extraordinarily rich catalogue of visual representations along the way. Another part is to ask why this organ has been so elevated, revered, celebrated and even petitioned. No other organ (and certainly not the brain) has commanded such pervasive and diverse attention; the heart is by far the most richly depicted of all the organs in the body. It is through the history of depictions of the heart that I hope to explore how it has been regarded and how that regard has, in turn, been intimately related to, and influenced by, the prevailing religious, cultural, artistic and scientific orthodoxies of the era in question.

Earliest depictions

In approaching the subject, I have been conscious of my inherent bias towards Western traditions, given my own background and exposure, and in particular my relative ignorance of Eastern and African art and medicine. It is certainly true that the heart has been depicted more widely and variously in Western cultural traditions than elsewhere, but in these pages the reader will nonetheless encounter depictions of the heart from China, India and pre-Columbian central America. Indeed, a fundamental premise of this book is that the enduring properties we attribute to the heart have been recognized almost universally, across time and place. At the same time, a lineage of evolution of ideas around the heart can be traced through art, religion, secular life and what came to be the world of science.

We have no *visual* evidence of how classical writers conceived of the heart. Only when Greek and Roman texts referring to the heart were rediscovered, reinterpreted and represented in derivative Western European works in the twelfth century CE were they accompanied by anatomical illustrations (see Chapter 2).

For the earliest (surviving) depictions of the heart we move to ancient Egypt, a culture with knowledge of the human body that was built on invasive exploration. The *Book of the Dead* is a modern term for various collections of ancient Egyptian funerary texts that first emerged in the seventeenth century BCE. The *Book of the Dead* comprised versions of texts (they do not exist in any canonical form) that were syntheses of religious and mythological beliefs, organized as an instruction manual, or book of spells, in chapters for particular purposes.[2] The *Book of the Dead* contained information to enable a dead person to navigate the complex route to the afterlife. It was recorded on the insides of coffins, on the walls of burial chambers and, for the most part, on thousands of rolls of often richly illustrated papyrus.

Again, in this tradition the heart was regarded as the functional core of a person and the site of the mind: the centre of thought, memory, will and intention. Accordingly, the heart contained the essence of the individual (there was no soul per se), which was itself both multifaceted and divisible. After death, the Egyptians believed a final judgement took place in the 'weighing of the heart' ceremony, in which the gods determined whether the deeds of the deceased while alive (symbolized by the heart) were in accordance with the moral order of the goddess Ma'at, adjudged by the weight of the deceased's heart against her feather. In Figure 1.1, Anubis, the jackal god of mummification, weighs the heart. On the other side of the scales is an ostrich feather, regarded as a symbol of truth. A favourable judgement, in which the balance stayed level, would allow the deceased to join Osiris in the afterlife. In an unfavourable case, the heart was consumed by the crocodile-headed monster-goddess Ammut, leaving the deceased trapped in *Duat*, the underworld.

Though depicted in literal form, with the heart and feather on opposing sides of a scale, the judgement ceremony was metaphorical. In the burial process, all the organs except the heart were removed from the body of the deceased and kept in canopic jars to accompany it. The embalmed heart was returned to the mummified body, since it was believed to be essential for use in the afterlife (unlike the pulverized brain, which was discarded, having been extracted through the nose with a tool not unlike a modern lobster pick).

Two ancient Egyptian words refer to the heart: *ib* and *haty*. The precise relation of each to the other is disputed. The *ib*, according to some, is the seat of emotions, individuality and will, while the *haty* corresponds to the anatomical organ. As part of the burial ceremony, a heart amulet was placed on the front of the chest to plead the case for the deceased and to limit the disclosure of any information by the heart that might prejudice safe passage to the afterlife. Heart amulets that were laid on the body often took the form of a scarab beetle – their purpose and composition (green jasper, serpentine, basalt, obsidian) being defined in Book 30 of the *Book of the Dead* and inscribed with hieroglyphs.[3, 4]

> Oh my heart [*ib*] of my mother
> Oh my heart [*haty*] of my mother
> Oh my heart of my being
> Do not rise up against me as witness
> Do not oppose me in the tribunal
> Before the great god, the lord of the west [Osiris]
> Lo, your uprightness brings vindication.

Thus, for the Egyptians, the heart was a data repository, a container, a record of the deeds of the deceased.

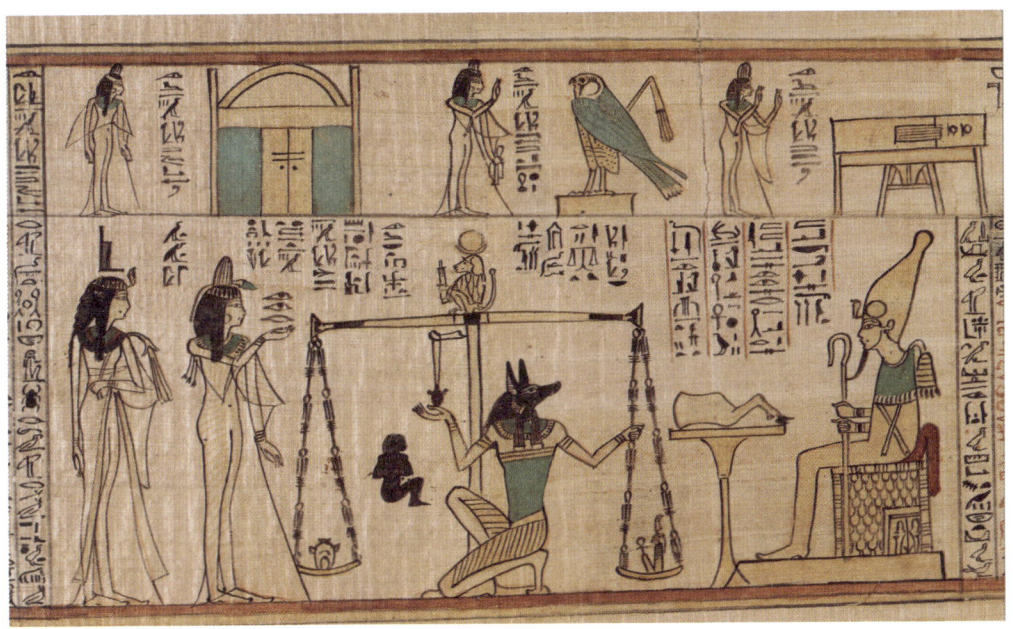

FIGURE 1.1 (top) *Book of the Dead*, Papyrus of Ani, 19th Dynasty, *c.* 1200 BCE, Egypt. (bottom) *Book of the Dead* for the Chantress of Amun Nauny, *c.* 1050 BCE, Egypt.

FIGURE 1.2 (top) Funerary papyrus for Djoser, a priest of the goddess Bastet; Ptolemaic period, *c.* 300 BCE, Egypt. (bottom) Weighing of the heart (detail).

Our own language extends this notion. The word 'record' is derived from the Old French word *record*, or 'remembrance', from *recorder*, meaning 'to bring to remembrance', which in turn comes from the Latin *recordari*, 'remembering', itself derived from *cor, cord* – 'heart'. As a noun, the word was first used in legal contexts to mean 'writing down in evidence'; as a verb, it meant 'to narrate orally or in writing' and also 'to repeat, so as to commit to memory'. Even today, we speak of learning 'by heart'.

Many versions of the *Book of the Dead* were illustrated. The sequence of the judgement ceremony and its critical characters and elements were depicted alongside scripted, formulaic declarations of the 'right deeds' of the deceased. While the elements were consistent, the actual depictions were not entirely stereotypical. The artists clearly had some latitude, including in the representation of our key element, the heart. In the example shown in Figure 1.1, the heart is anatomically isolated and presented as a self-contained unit. Here, it is rust-red – though possibly it may once have been more vividly coloured – and balanced upright on its apex, with the severed major arteries and veins of the base tied with ligatures and pointing upwards. In other versions, the deceased is depicted carrying his own heart towards the scales. Still other examples show the heart symbolically, as an urn or a vessel, with no attempt at anatomical realism (Figure 1.2). The heart is, of course, a hollow organ, and so this idea of the heart as a vessel will appear again and again.

The Egyptian context was addressed specifically in the 1920s in the specialist monthly journal *Regnabit*. This French Catholic devotional publication had the singular purpose of publishing articles on religious iconography in relation to the heart. Until it fell out of favour with the church, it seems to have embraced articles on comparative religious iconography. Contributors included the specialist in Christian symbolism, iconographer and engraver Louis Charbonneau-Lassay (1878–1946) and his associate, the scholar and philosopher René Guénon (1886–1951). Writing in 1925, Charbonneau-Lassay noted:

> In their [ancient Egyptian] hieroglyphs, sacred writing wherein the image of the thing itself often represents the very word that designates it, the heart was nonetheless depicted only by an emblem: the vase. Is not the heart of man in fact the vase in which his life is continuously maintained by means of blood? [5, 6]

Elsewhere, Guénon develops ideas around the heart as the vessel for Christ's blood, including in relation to the Holy Grail, the cup from which Christ drank at the Last Supper and which, according to legend, was brought to England by Joseph of Arimathea. It is remarkable how often the theme recurs, with the heart emerging not only as a vessel to contain the characteristics of the individual, but as a receptacle for occupation and habitation by God or gods.

The Aztecs believed that the heart (*tona*) was both the seat of the individual and a fragment of the sun's heat (*istli*), and the sun deity itself was considered a heart-soul: 'round, hot, pulsating'.[7] The divine fragment was encased within the body and was vulnerable to corruption by humans. Conversely, heart extraction was believed to liberate the *istli* and reunite it with the sun. The sixteenth-century Aztec Codex Magliabechiano Folio 70 (Figure 1.3) documents the ceremonies of the Aztecs in which human hearts were excised, still beating, from the living, and delivered to the sun.[8] To access the heart, a cut was made in the abdomen and from there through the muscular diaphragm into the chest. The figure shows the act of sacrifice, with a restrained man supine on a flat altar. He is held by the legs while the priest, or other operator, incises the abdomen. The heart, partially filled with blood, is released and the *istli* returns to the sun, leaving a trail of blood in its wake. Several such sacrifices were made in the same ceremony. At the base of the temple structure, an earlier heart donor with gaping wounds is seen being dragged away by the arm.

Ancient Indian oral traditions recorded in the Upanishads (*c.* 800–500 BCE) attribute diverse properties to the heart. These texts, presented in thematic clusters of verses, provided expansive illumination of spiritual, philosophical and practical matters. Significantly, they are not divine texts. These verses are not revelation; they are not considered the 'word of god', but to derive from enlightened human insight and deep intuition. In light of this, the way in which the heart is depicted in these texts should be instructive. *Upanishad* is derived from the Sanskrit for 'to sit close' and implies the original mode of transmission, by oral means, from sage to student. A central theme of the Upanishads is the relationship of humanity and the universe, and the unity of *ātman* ('the soul') and *Brahman* ('the ultimate reality'). Ideas in relation to the heart figure prominently and are considered in detail by others.[9] However, I am drawn to the threads held in common with other cultures: the heart as seat of self; source of knowledge, emotion and life origination; and interconnected centre, relating to the branching vessel structures to which it is attached. These ideas resonate with elements of heart symbolism from subsequent Western traditions. The Upanishads make intriguing statements about the heart as the originator of the life force, both in the universal sense and at the human level.

The Brhad-āranyaka Upanishad (V.6.1) says:

> This person who consists of mind is of the nature of light, is within the heart like a grain of rice or barley. He is the ruler of all, the lord and governs whatever there is.[10]

FIGURE 1.3 Ceremonial rites, Aztec Codex Magliabechiano; panel 70R, mid-sixteenth century, Mexico.

The Chandogya Upanishad (III.14.2–4) likewise locates the *ātman* (or true self) within the heart in the repeated refrain:

This is myself within the heart, smaller than a grain of rice, than a barleycorn, than a mustard seed, than a grain of millet or than the kernel of a grain of millet. This is myself within the heart, greater than the earth, greater than the atmosphere, greater than the sky, greater than those worlds…
Containing all works, containing all desires, containing all scents, containing all tastes, encompassing this whole world, without speech without concern: this is the self of mine within the heart; this is Brahman.[11]

Verses elsewhere in the Brhad-āranyaka Upanishad (III.9.19–25) refer to the heart 'supporting' the functions of hearing, sight, speech, initiation, faith and truth, and sexual desire. Intriguingly, the heart was also proposed as the source of semen:

On what is semen supported? On the heart he [Yajnavalkya, a Vedic sage] said. Therefore, they say of a newborn who resembles his father that he seems as though he has slipped out of the heart, he is built out of his heart; for on the heart alone is semen supported.

As far-fetched as this seems, bearing in mind our contemporary knowledge of the origin of semen, almost identical ideas can be found in Western beliefs; later, we will encounter an altogether imagined anatomical correlate, in which the heart was 'plumbed' to the penis, and which was even depicted by Leonardo da Vinci as recently as the end of the fifteenth century.

The nineteenth-century French author and scholar of iconography René Guénon considered the heart in his book *Symboles de la Science*

Sacrée (*Fundamental Symbols: The Universal Language of Sacred Science*).[12] Noting the references to seeds within the heart in passages from the Upanishads, he also drew the reader's attention to a seventeenth-century print, contained within an academic thesis, in which the Hebrew letter *yod* had been inscribed three times in the heart of Christ. The letter was used to represent what Guénon termed the 'Principle' and it appeared in triplicate to signify the Christian Holy Trinity. The significance of this from our perspective is the decision to place the script not in the brain or deep in the abdomen, but in the heart of Christ.[13] Guénon comments:

> The yod in the heart is therefore the Principle residing at the centre, be it from the macrocosmic point of view at the 'centre of the world' which is the 'Holy Palace' of the 'Kabbala' or from the microcosmic point of view in every being, virtually at least, at his centre, which is always symbolised by the heart in the different traditional doctrines, and which is man's innermost point, the point of contact with the Divine.

He goes on to say that *yod* has also been used to denote a seed, a direct echo of the passages from the Upanishads quoted above. Remarkably, a similar notion crops up in the black grain or seed (*granum nigrum*) depicted in the hearts of twelfth-century Western anatomical drawings. Since these seeds do not reflect anatomical reality – the human heart contains no such seed-like form – it is remarkable that they should occur, with shared connotations, in two different cultural traditions.

So, in common with early concepts in both Egypt and Greece, there was in India a seamless amalgamation of the spiritual and the physical. There were notions of the heart as a vessel, a shelter, an originator of life, the source of semen, the seat of sentience and the home of the soul.

The town of Chidambaram, in the southern Indian state of Tamil Nadu, is the site of one of the five principal temples to the god Shiva. Here Shiva is manifest as Nataraja, Lord of Dance. As Lord of Dance,

Shiva performs the Änanda Thaandavam ('Dance of Bliss'), the dance in which the universe is created, maintained and dissolved. The architecture of the temple itself is laden with anthropomorphic allusions, from the nine entrances denoting the nine bodily orifices to the 21,600 gold-covered copper sheets that make up the roof of the inner temple, or *sanctum sanctorum*, reflecting the number of breaths a person takes each day. On the spot where Shiva is said to have performed the Änanda Thaandavam is the *Ponnambalam* (*Pon* meaning 'gold', *Ambalam* meaning 'stage'), housing the iconic statue of Shiva in his dancing form. The *Ponnambalam* is at the core of the temple, but it is intentionally offset from the centre of the temple, as the heart is in the body. The cosmic dance of Shiva at Chidambaram is conceived as analogous to the human heart.

We will see how the conception of the heart as the dwelling place of Christ became prominent in Christian theology in the Middle Ages. Similarly, the Hindu epic of the *Ramayana* (*c.* seventh-century BCE to third-century CE) refers to the monkey king Hanuman – a devotee of the god Ram and regarded as a minor deity himself – ripping open his chest in an act of supreme devotion to reveal Ram and Sita dwelling within his heart (Figure 1.4 and 1.5). There is some debate as to whether the authentic reference is to Hanuman revealing Ram and Sita in his heart or the names Ram inscribed on each of his bones (ribs), but Hanuman is often shown in paintings and religious figurines sitting, crouching or standing, clasping the edges of his open sternum in each hand to reveal, through a central elliptical window, the figures of Ram and Sita – pristine, elegant and decorated – within. As in the Kalighat-style images here (Figures 1.4 and 1.5), the anatomical heart itself is not represented. The flesh wounds can be more or less gorily depicted: clean-edged as in these versions, or with floridly bloodied margins in others, where the deities are within the bleeding heart, drawing obvious comparisons with later Christian traditions, and in particular the Catholic Cult of the Sacred Heart, which we will look at in detail in Chapter 6.

Visual representations of hearts in Indian 'medical' traditions, such as they are, do not attempt realism; they are not only non-anatomical, they are often infused with rich metaphor inspired by the natural world, with an emphasis on charting the chakras (which can be considered as energy centres in the body) and metaphysical energies.[14]

The Chandogya Upanishad contains the following verses:

Within the city of Brahman, which is the body, there is the heart, and within the heart there is a small dwelling. This has the shape of a lotus, and within it dwells that which is to be sought after, inquired about, and realized. What then is that which, dwelling within this little house, this lotus of the heart, is to be sought after, inquired about, and realized?

Chandogya Upanishad (VIII.1.1.2)

As large as the universe outside, even so large is the universe within the lotus of the heart. Within it are heaven and earth, the sun, the moon, the lightning, and all the stars. What is in the macrocosm is in this microcosm.

Chandogya Upanishad (VIII.1.3)

These brief verses are remarkable in the scope of the properties they attribute to the heart. They describe the localization of the soul in the heart, articulate the universality principle of Vedanta (the oneness of *ātman* and *Brahman*) and allude to the body as a microcosm of the universe. The reference to the lotus flower becomes a starting point for depictions of the heart, in which both physiological and psychic

FIGURE 1.4 (overleaf left) Hanuman revealing Rama and Sita in his heart, nineteenth century, India.

FIGURE 1.5 (overleaf right) Hanuman revealing Rama and Sita in his heart, *c.* 1830, India.

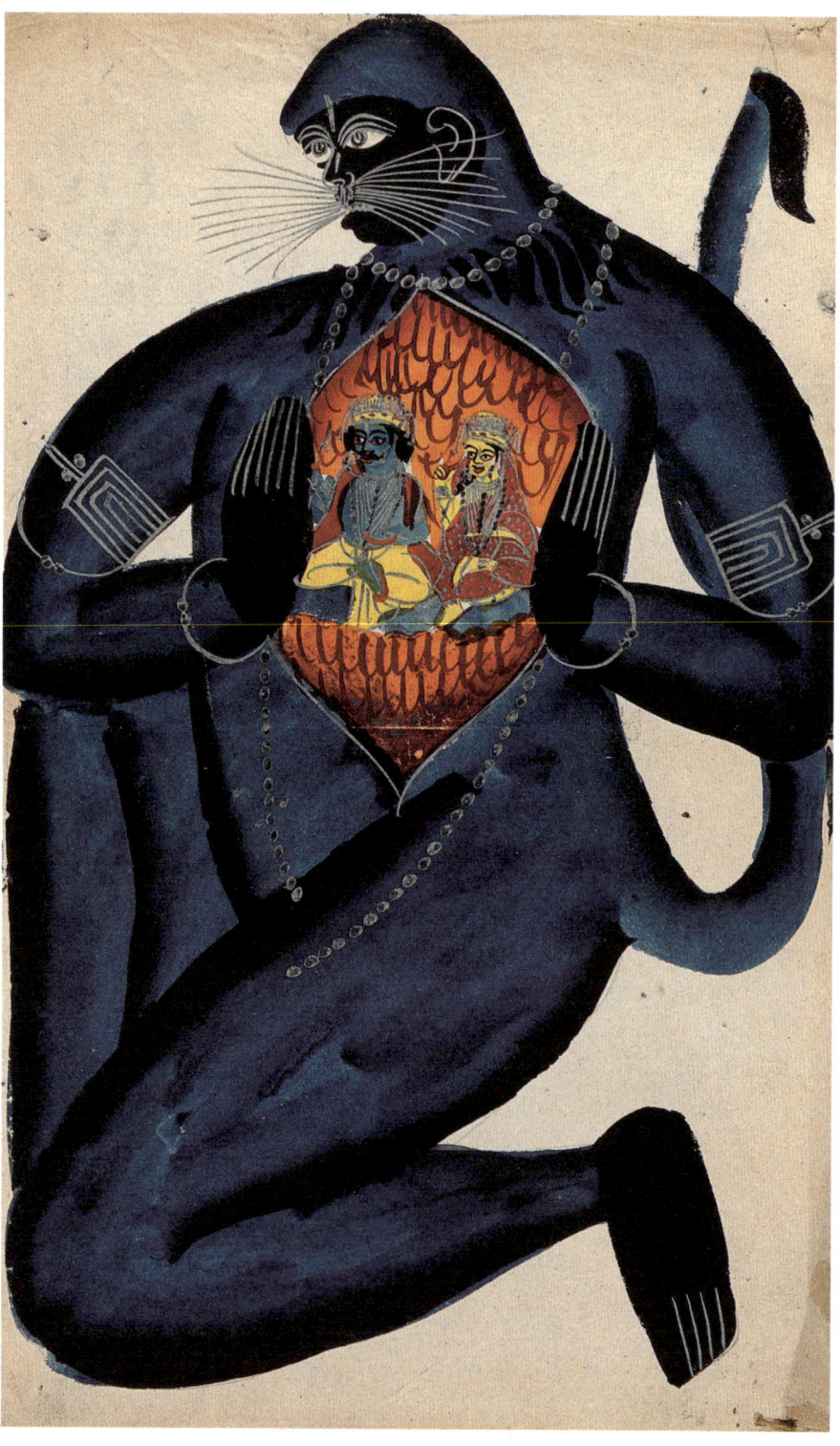

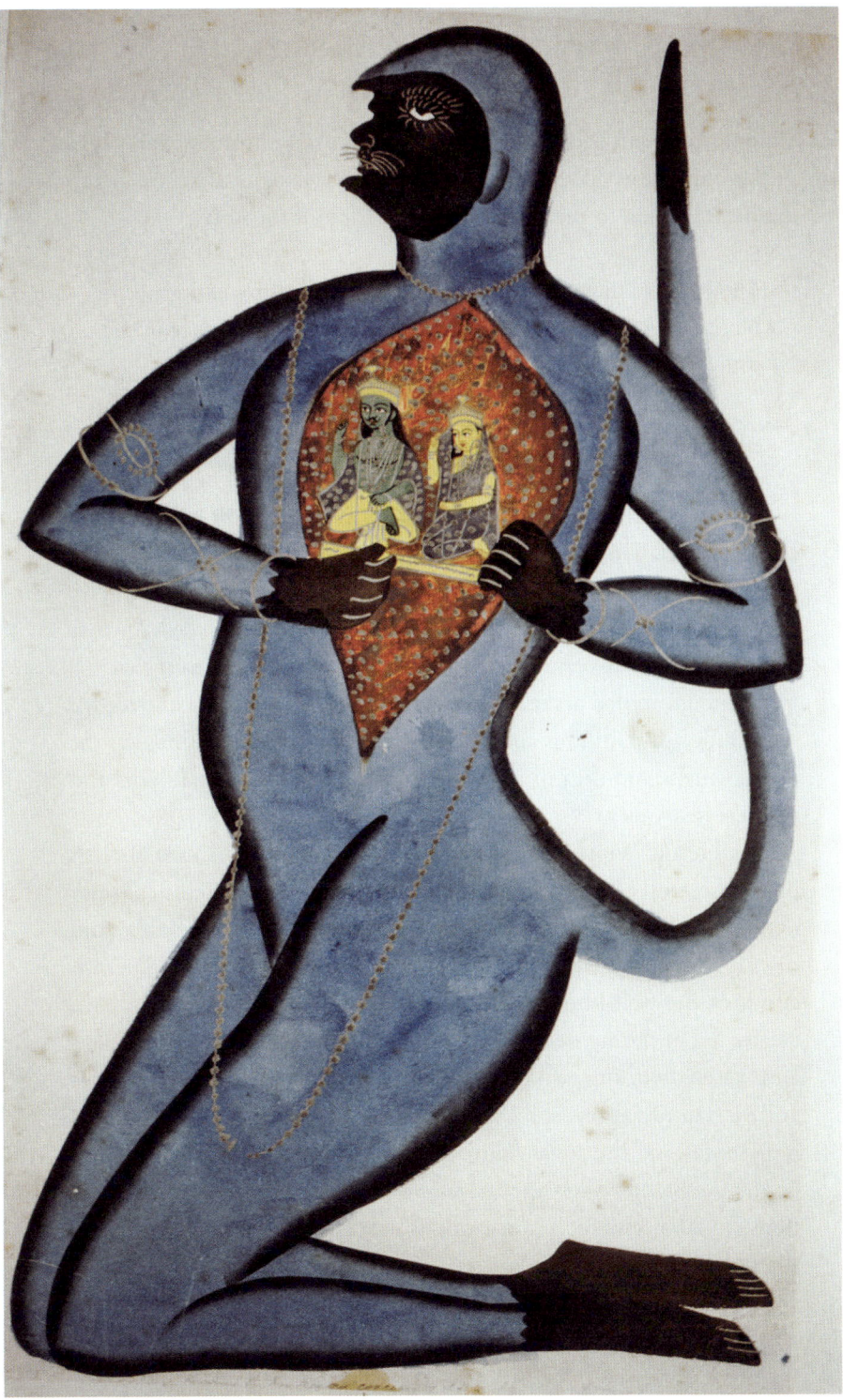

connotations can be shown through the use of colour and variations in the number of petals.[15] The opening and closing of the lotus petals has been likened to the phases of contraction of the heart.

> The heart is like a lotus facing downward. On waking up, it blooms; on sleeping it closes up. That is the resting place of the soul. It is the supreme location of the consciousness.[16]

In the relatively recent (early twentieth-century) Buddhist painting, or *tangka*, shown in Figure 1.6, the heart of the lower left figure is shown as a lotus flower. The image is originally derived from seventeenth-century Tibet, which saw a blossoming in medical interest and even the construction of a monastic medical college. A set of seventy-nine *tangkas* illustrated a comprehensive four-volume medical treatise called *The Blue Beryl*. Created between 1687 and 1703, the original paintings were a form of educational art that interweaved practical medical knowledge with Buddhist traditions and Tibetan lore, depicting such things as the use of omens and dreams for making diagnoses, medicinal herbs and medical instruments.

Sir Henry Wellcome (1853–1936), the American-born British pharmaceutical entrepreneur and philanthropist, was a prolific collector of anthropological and medical artefacts, which are now held by (and displayed at) the Wellcome Collection in London. The collection contains the well-known *Ayurvedic Man*, which is thought to have originated in the eighteenth century in Nepal (Figure 1.7).[17] This pen-and-watercolour image hints at the presence of vessels, bodily cavities and viscera of various sorts, but at its core this is not an anatomical representation. The squat figure with splayed limbs, staring eyes and branching blood vessels appears superficially similar to the anatomical drawings that emerged in Europe in the twelfth century, and must surely be derived from them (see Chapter 2), but the internal organs appear more like an array of diodes in an electrical circuit than a realistic anatomical impression. It would be more appropriate to describe it as a presentation of an alternative system of knowledge and beliefs that may have originated from a collaborative process between a physician, who was a scholar of Ayurveda, one or more artists, possibly from Kathmandu, and a calligrapher, who copied the texts.[18]

Given its long history of medical and, more particularly, medicinal practice and the ancient tradition of printmaking, one might imagine that Chinese scholarly texts would contain references to the heart, accompanied by illustrations. To some extent they do, but the pictorial medical record is dominated by images of commodities, especially plants, with medicinal attributes. Medicinal texts tended to contain pictorial albums, sometimes with hundreds of polychrome paintings depicting the morphology of medicinal substances.[19] By the Eastern Han period (25–220 CE), there had been many attempts to systematize medical theory, and consensus was emerging about the practices of acupuncture and moxibustion. Texts from this period record networks of 'channels' and locations for the application of needles and the herb moxa (termed '*moxibustion*'), and various surgical-style instruments for treatment. In 1972, medical manuscripts dating to the Eastern Han dynasty were discovered at Hantanpo in modern Wuwei prefecture, Gansu province.[20] They are the earliest known texts describing the therapeutic use of fine needles at specific points on the body. They also illustrate ideas about the movements of spirits in the body, and the need to protect their free flow. In 1973, silk manuscripts excavated from the Mawangdui tomb site in modern Changsha, Hunan province, provided new evidence from the earlier Western Han dynasty of a system involving eleven 'channels' of energy flow in the body, and a method of taking the pulse at various parts of the body to diagnose the condition of those 'channels'.[21]

The emphasis is on the depiction of a system, rather than a literal description that itemizes the physical components and their spatial relations. The anatomical connections of, for instance, the pulmonary

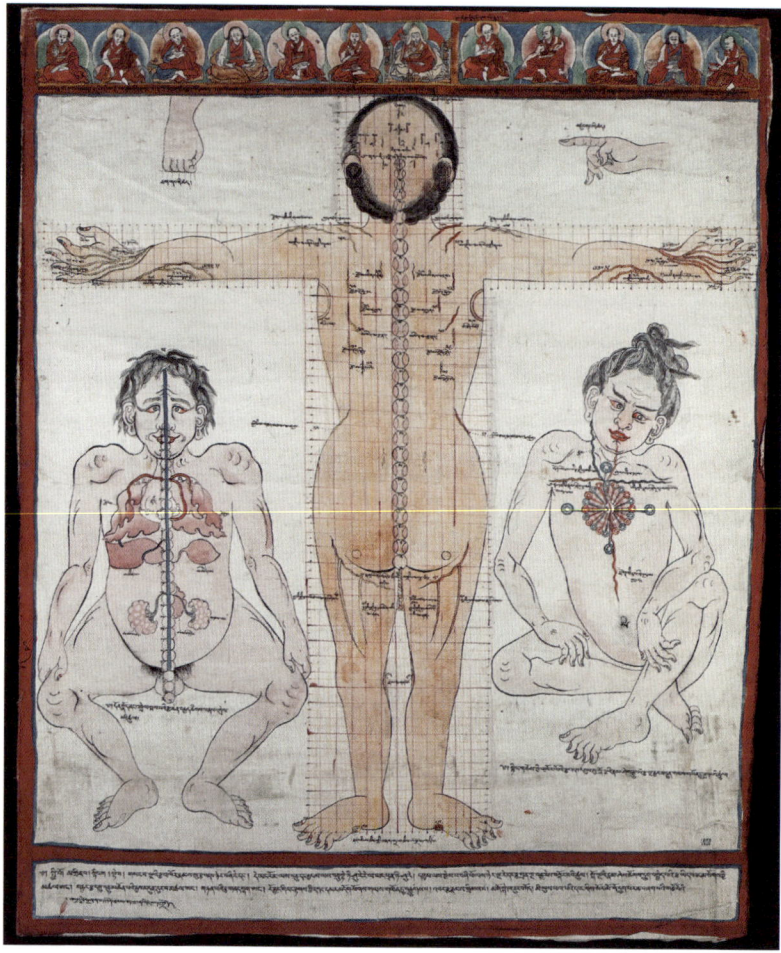

FIGURE 1.6 Anatomical figures, early twentieth century, Tibet.

artery or the inferior vena cava were not important in this paradigm since they did not matter to the overall framework of understanding (Figure 1.8). Even attempting to interpret Chinese medical practice by regarding visual representation through a conventional Western interpretive lens is likely to result in a blurred appreciation and risks serious misinterpretation of the Chinese method.[22]

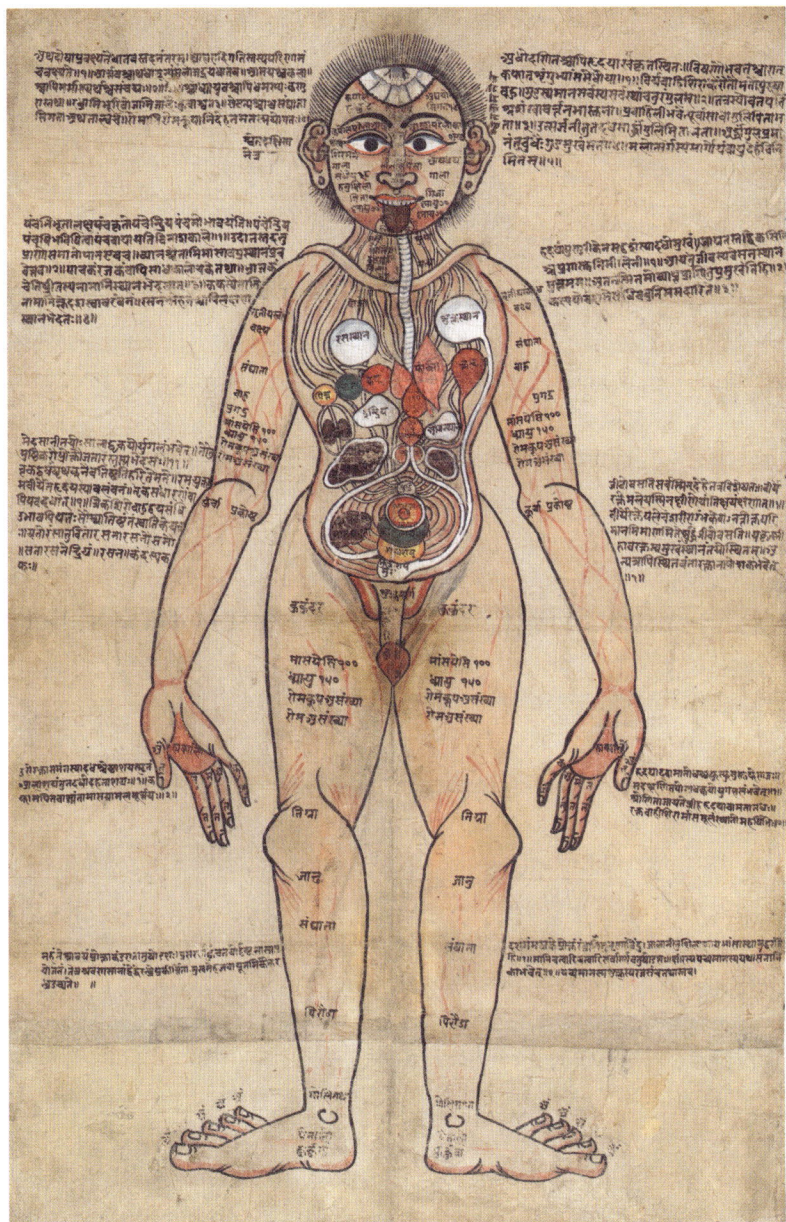

FIGURE 1.7 *Ayurvedic Man*, eighteenth century, Nepal.

Nonetheless, attempts have been made to expose Chinese medical syntheses to the West. The earliest Western depictions of the heart that were derived from Chinese anatomy came from Andreas Cleyer, a German physician of the Dutch East India Company, in his *Specimen medicinae sinicae* (*A Look at the Medicine of China*), published in Frankfurt in 1682. He worked with the Jesuit Michael Boym, known for his studies in botany and his descriptions of Chinese pulse lore. Cleyer's images of the heart (Figures 1.9 and 1.10) form part of a collection of woodcuts that include references both to acupuncture and to detection and interpretation of the pulse. Unlike almost any other early image, the body is shown in the so-called sagittal cross-section, as though it has been cut down the centre between the eyes and dividing the nose. Rings of cartilage encircle the trachea, which descends directly into the heart. The acorn-like heart itself sprouts a leash of vessels that are plumbed into the viscera.

The text on the images is in Latin. The label adjacent to the double layers around the heart is *involucrum cordis*, the 'envelope of the heart'. Ludwig Choulant, a nineteenth-century historian of anatomical depictions, expands further:

> Through the agency of Dr G Schultz, prosector [a person involved in the preparation of dissected specimens] at St Petersburg, four genuine Chinese plates acquired by the Russian mission in Peking were obtained for the library of the Medico-Chirurgical Academy and since they are beyond any doubt originals, an exact description of the illustration follows...
> The first plate contains the profile view of the trunk and head of a human body without the arms, legs and genitals. It resembles the figure represented on the first of Cleyer's plates...
> On the figures themselves is a great deal of Chinese writing in various sizes; a small part was translated by the well-known missionary Gutzlaff... during a brief stay in St Petersburg...[23]

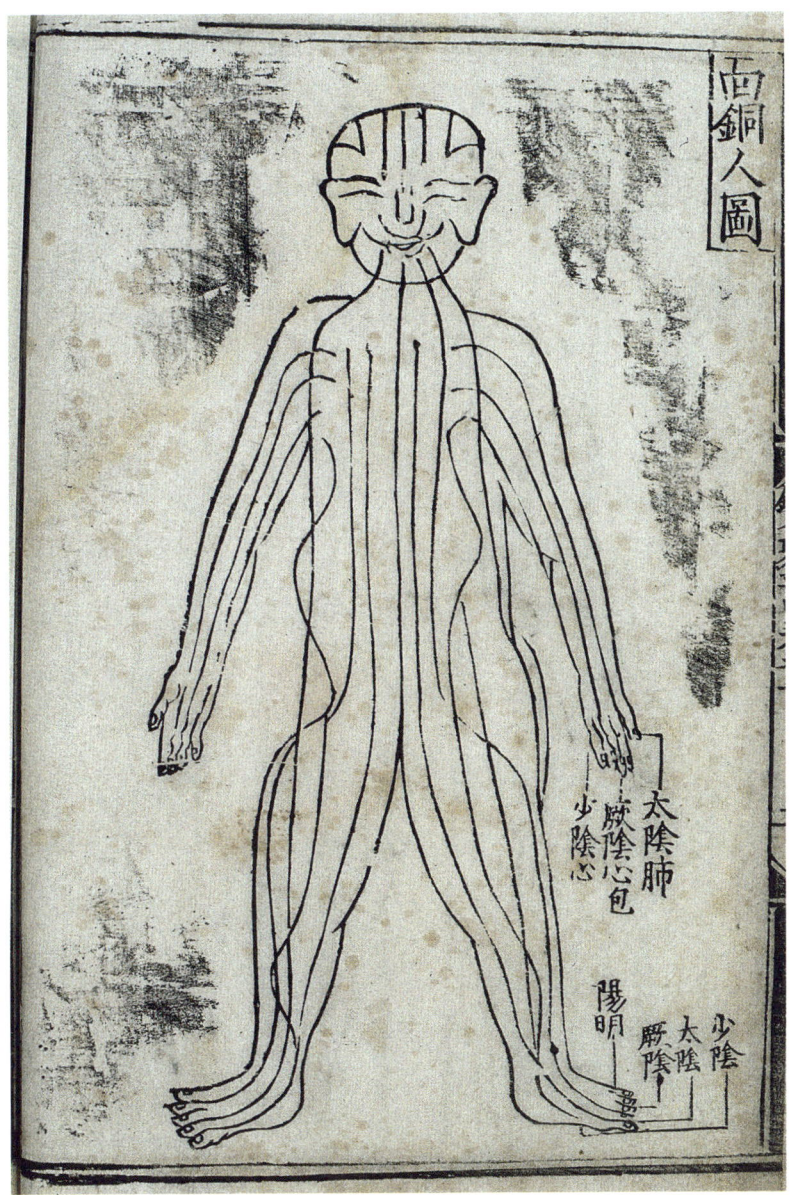

FIGURE 1.8 Standing Bronze Man, Zhenjiu juying 針灸聚英 (*A Glorious Anthology of Acupuncture and Moxibustion*, 1519). This edition from 1537, China.

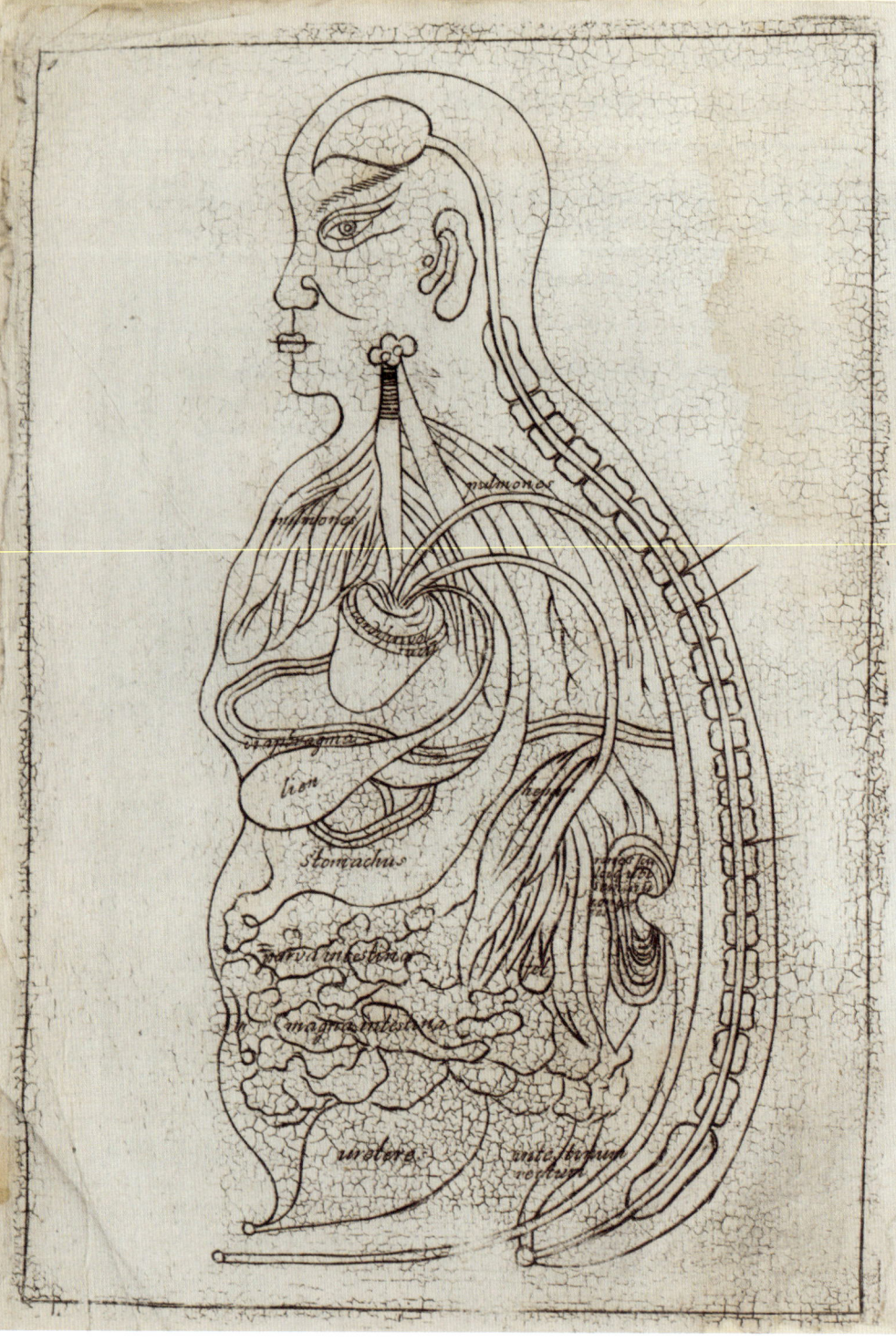

Gutzlaff's translation begins:

The heart is the monarch of the whole body. The lung is the communicating principle which rules over all the members (prime minister). The liver is so to speak the general for the strategic branch.[24]

For all the divergence between the narrative systems of Chinese and Western medicine, it is telling that each places the heart at the centre of operations. By likening the heart to the king, the author of the Chinese text shares an intuition that was articulated by many others. William Harvey uses the same metaphor in his dedication of his great work *De Motu Cordis* (*On the Motion of the Heart*) to Charles I of England, while countless artists from Christian traditions have depicted the heart of Christ beneath a crown of thorns. We will explore each of these in later chapters.

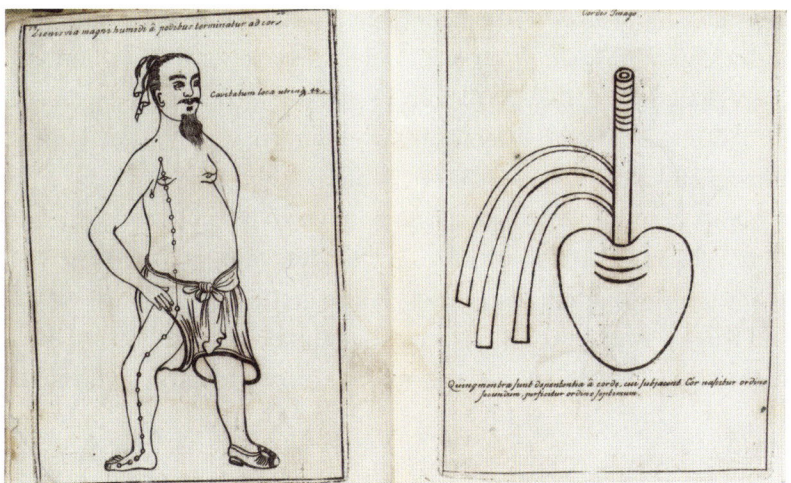

FIGURE 1.9 (opposite) Andreas Cleyer, *Specimen medicinae sinicae*, Frankfurt, 1682.
FIGURE 1.10 (above) Andreas Cleyer, *Specimen medicinae sinicae*, Frankfurt, 1682.

The remarkable fact that this and other ideas of the heart and its attributes are held in common across time and different cultures is the starting point for my exploration. Could it be that our universal, shared experience and appreciation of the beating and responsiveness of our own hearts makes this conceptual convergence almost inevitable? After all, how we understand or interpret the heart's properties is, in part, determined by what we see and feel. Even so, the concordance between Eastern and Western traditions and thinking in respect to specific attributes of the heart is striking – and all the more so because these properties seem to be intuited, rather than reasoned, and because the recurring themes or clusters of ideas are often distinctly contained and not intrinsically related to each other.

A significant outlier is the apparent absence of depictions of the heart in African traditions, though in an historical cultural context, even the term 'African' is a largely imposed European notion.[25] The creative endeavours of the thousands of peoples, tribes and communities of that continent were seemingly not focused on depicting the inner workings of the individual human body. They give more emphasis to ancestry; devotion and ceremony; divination, sexuality and fertility; personal decoration and the ornamentation of objects of use. Even in these contexts, depiction of the heart does not seem to figure in any prominent way in the traditional 'artistic' endeavours of sub-Saharan Africa.

It has been suggested that the near-universal perception of the heart comes close to Carl Gustav Jung's (1875–1961) theory of the archetype as an enduring pattern that, although manifesting in specific historical and religious cultures, has an ever-present origin that Jung traced back to the psyche, or in his words, the collective nature of the unconscious.[26]

[the] Jungian hermeneutic which, building upon his archetype theory, posits that similar psychological patterning and functions exist in specific religious symbols across varying traditions—in this case, that of the heart.

Psychologically, Jung likened the impersonal awareness that arises in the heart with the conscious formation of the Self. 'That is the first inkling of a being within your psychological or psychical existence that is not yourself—a being in which you are contained, which is greater and more important than you but which has an entirely psychical existence'. [27]

However, in exploring how and why we perceive and depict the heart as we do, there is no need to invoke some mysterious 'collective unconscious' to explain the universally attributed properties of the heart. We can approach many ideas that have grown up around the heart and its supposed functions based on a *common conscious experience*. We have a shared, but also highly personal, experience of our hearts that is declared in a broad and eclectic range of visual depictions.

In visiting the various representations of the heart, we need not cling to intellectually enticing strands of continuity, though there are many. That would be to miss the point. It does not matter if the multiple ways in which we choose to portray the heart have some identifiable ancestral lineage. They might do, or they might not. The point is that these ideas have been generated, retained, propagated and embellished across many cultures separated by time and space.

If human awareness of this seemingly sentient organ affords it such wide cultural relevance, a fundamental consideration is its relation to its 'host'. The heart beats spontaneously; neither its rate nor its force of contraction is under conscious control, and yet we can be aware of changes in each. This autonomy of the heart – that it seems, in some very tangible sense, to *function apart from and communicate with our conscious selves* – underlies the properties with which we imbue it and, as a consequence, its unique position in art, literature and religion.

The Medieval Heart

*Lo spirito della vita, lo quale dimora nella
segretissima camera del cuore.*
('*The spirit of life which dwells within the most
secret chamber of the heart.*')

DANTE ALIGHIERI
Vita Nuova, 1294

The early medical heart

In searching for explanations for difficult problems, our species tends to seek to impose coherent narratives; we take what is known (or believed) and then 'join the dots' with speculation and conjecture until a tractable story emerges. This is not the preamble to some remote historic anecdote – we do it now. Large parts of current scientific thought and endeavour are pasted onto sometimes quite flimsy scaffolds. The better these narratives explain what we are trying to grasp, the fewer the inconsistencies or contradictions, the better the fit with prevailing thought, the more likely that the theory and its accompanying narrative will persist. It can often seem that the 'fit' matters more than whether the story is right. Whatever the benefits, such narratives can ultimately constrain fuller understanding.

If we look back some two thousand years to the second century CE, we find that the physician and philosopher Galen (*c.* 129–210 CE) devised an outstanding narrative: a synthesis for human 'systems biology' that was sufficiently good to persist as the central theory of medical 'sciences' for more than 1,300 years. Galen was born in Pergamon, now in modern Turkey, and worked in Rome, serving as personal physician to several emperors, including Marcus Aurelius. His works became a written reference and a repository of the thinking up to his own time, but they also contained his own ideas and syntheses, particularly in relation to the functioning of organ systems. Since the laws of the time prohibited the dissection of humans, his theories were derived in part from animal dissection and in part from the conventional thinking of the day. He drew selectively from the work of Hippocrates and incorporated in his thinking the theory of the four humours of Hippocratic medicine: black bile, yellow bile, phlegm and blood. According to this tradition, when the humours were in balance, the body was healthy. But Galen also moved to a more functional consideration of the organ systems and their interrelations. Indeed, his principal anatomical treatise, *De usu partium*, can be literally translated as '*On the Usefulness of the Parts*'. He presents each organ system and their components as fulfilling a larger role within the operation of the body, discussing interactions with the broader physical and spiritual world, and directing his writing towards both philosophers and physicians.[1]

Here, we are most interested in his theories in relation to the heart and the blood. We should pause for a moment to consider Galen's synthesis, because it was so influential that it serves as a reference, and because it will become a pivot point for the transitions to modern medical thought. For our purposes, Galen's key points were the following:

- Blood was made in the liver and received its nutritive properties from the intestine.
- Blood was passed to the heart, where it mixed with *pneuma* (breath) from the lungs.
- This 'vivification' of the blood was accomplished by the passage of blood through invisible pores in the septum, which divided the left and right ventricles of the heart. I emphasize the reliance on these pores since questions of their existence will recur in subsequent discussions around the work of both Leonardo da Vinci and William Harvey.
- Galen recognized the existence of systems of veins and arteries (veins that pulsed) but believed in a centrifugal model in which blood, which had been made and processed centrally, ebbed slowly to the peripheries, *where it was consumed*. The veins and arteries, in this model, would have necessarily contained blood of different origins and compositions, which they delivered separately to the tissues.

I must stress that there was no concept of the *circulation* of blood. However, Galen's model provided a rational underpinning for the practice of bloodletting to balance the humours – one of the few medical interventions available at that time. For Galen it made perfect sense to remove bad blood, since the liver could then be induced to replace it with new, good blood, thereby restoring the patient to health.

Reproduced from Sir William Osler's *The Evolution of Modern Medicine* (1921), Figure 2.1 shows a nineteenth-century interpretation of the vascular systems according to Galen's description.[2] Note how the arrows indicate the direction of the ebbing of blood. The blood originating from the liver (marked 'C'), shown in black, converges in the peripheries with blood from the heart (with pores, marked 'P'), shown in white. There is no route back from the tissues, as there is no circulation. The image brings together Galen's synthesis with a modern style of portrayal in an anachronistic pseudo-anatomy. Whether Galen attempted any type of diagrammatic representation is not known; it is unlikely that he did, and certainly none survives. The first known depictions of Galen's classical descriptions come from twelfth-century Europe.

The Greek tradition in medicine had dwindled through the Middle Ages, when religious preoccupations quenched those of early 'science'. In fact, my reference to 'science' is a loose one, since the concept in its modern sense did not exist in medieval times. I am referring to attempts to understand the world through observation and, sometimes, experimentation. Ancient Greek and Roman traditions of philosophy and science, especially the works of Aristotle, were revived during the twelfth-century Renaissance through increased contact with the Islamic worlds of Iberia, southern Italy and the Middle East. The ancient texts were translated into Latin and entered Western Europe through such centres of scholarship as Salerno in Italy and Toledo and Cordova in Spain. In this way, Galenic ideas resurfaced in the High Middle

FIGURE 2.1 Charles Richet, Interpretation of the vascular systems according to Galen, 1879, Paris.

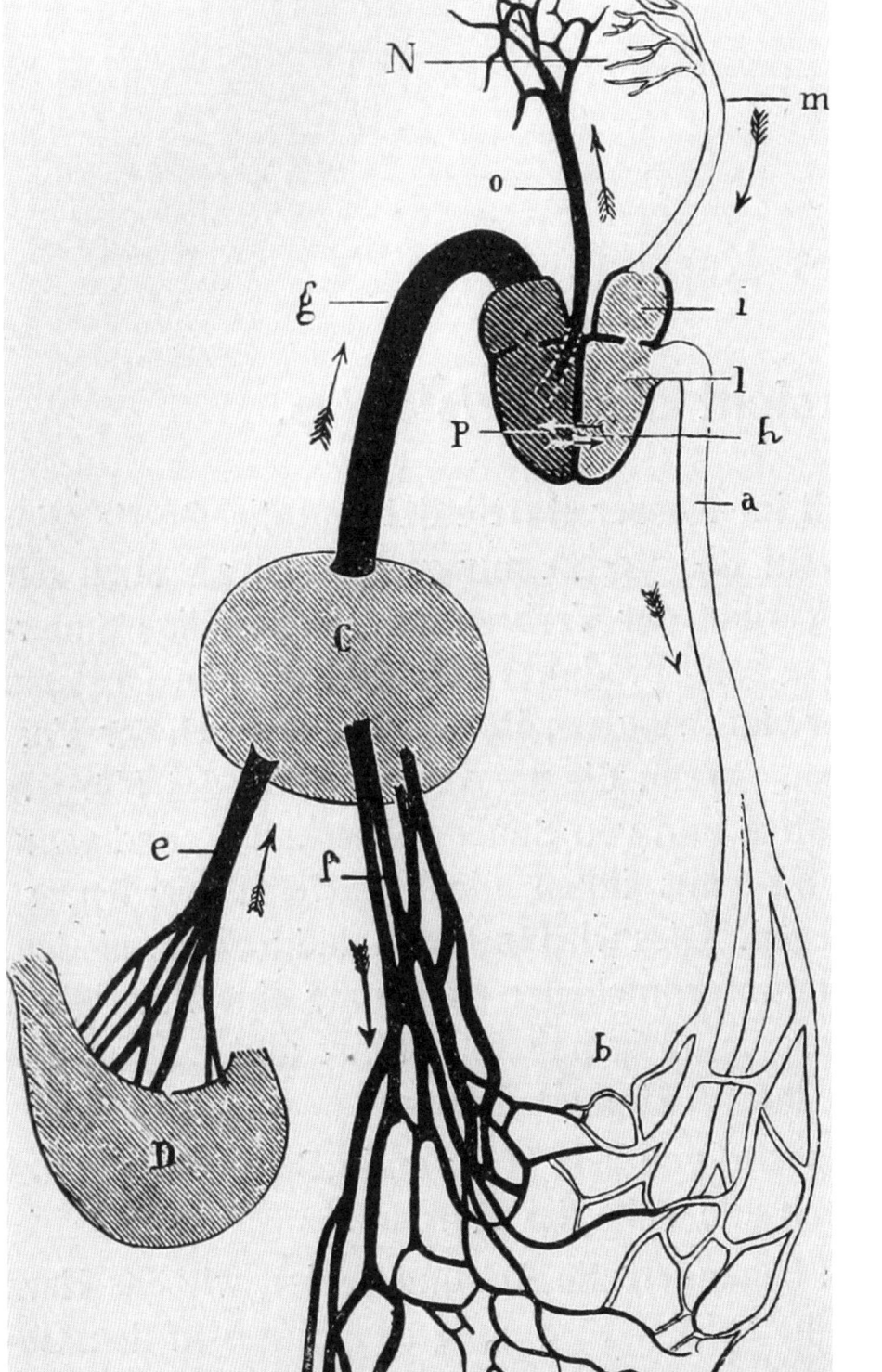

Ages, reaching Latin scholars in the West in the twelfth and thirteenth centuries, having been organized and rationalized by Arabic scholars, notably the philosopher-physician Avicenna (980–1037 CE), the pre-eminent medical scholar of his time, who codified medical 'knowledge' in his highly influential *Canon*. Whether there were contemporaneous Arabic pictorial depictions (now lost) or whether the Latin scholars were the first to make such depictions is obscure.[3] However, the earliest known anatomical pictorial depictions are the *Five-Figure Series* – reflecting Galen's systems of veins, arteries, bones, nerves and muscles – which was extended into a nine-figure series by addition of the male and female reproductive systems, some version of the alimentary system and the brain and ocular systems.[4] We have no record of the original artist(s), but the oldest surviving version, from the twelfth century, is in the library of Gonville and Caius College, Cambridge (Figure 2.2), with a later, more elaborate version in the Bodleian Library in Oxford (Figures 2.3 and 2.4). Again, we do not know whether the Oxford version was a direct or indirect copy of the Cambridge images, or whether they simply shared a common origin, but their stereotypical features strongly suggest that they originated (copied by hand, of course) from some common source, the identity of which is unknown.

The *Five-Figure Series* projects each of the individual 'systems' onto a template. The images themselves occupy almost the whole page and comprise the outline of a squat, seemingly flattened human form defined by a simple line. The open aspect and splayed limbs, and the deliberately convergent gaze, have been likened to the appearance of a dissected cadaver, though the images are not believed to be directly derived from actual human dissection. The organs are positioned asymmetrically in the torso and coloured brown. The liver is in the anatomical lower right (image lower left) and, consistent with Galen's model, is the source of the blood and the blood vessels. Counterintuitively, and contrary to modern convention, the 'arteries' are shown in blue and the 'veins' in red. While Galenic teaching did not define a precise functional delineation between these vascular systems, the description of arteries as 'veins that pulse' clearly implies some appreciation of a distinction between the two.

While the images provide diagrammatic representation of elements for each of the 'systems', there seems to be little or no attempt to portray an anatomical reality. The emphasis is on an overview of the perceived interconnections of the components in each of the systems and the basis is Galen's writings.

But even in these early works, the arteries and veins are drawn as separate systems. Knowledge of this separation can only have come about by open dissection – though not necessarily of humans; the received wisdom is that Galen did not dissect or have access to human specimens, and neither was morbid human anatomy practised in Europe during this period of revival of Galenic medicine.

The heart is shown just off the midline; it has the shape of an apostrophe and seems to be a node of some sort, from which the blood vessels emanate as fronded, rust-red strands (Figure 2.2). Intriguingly, in the centre of the heart is a small white spot that, on closer inspection, is the ultimate departure point of the vessels. There is no attempt at anatomical realism; this a schematic depiction of a written text. The absence of anatomical fidelity is emphasized by an accompanying diagram of the isolated internal organs (not shown), which depicts the heart as a simple inverted tear-shaped structure, again rust-red, with small lateral appendages and the central white spot.

The manuscript is an early example of graphic images as part of the substance of the work, rather than merely decorative illuminations. It is likely that these copies were made in monasteries, and some have speculated that they were manuals or guides for surgery.[5] But surgery is, above all, an exercise in applied anatomy, and it is hard to see what operation could possibly have been planned from these images, since they impart no meaningful anatomical reality. The text surrounding the figures, for the most part, does not relate directly to the figures

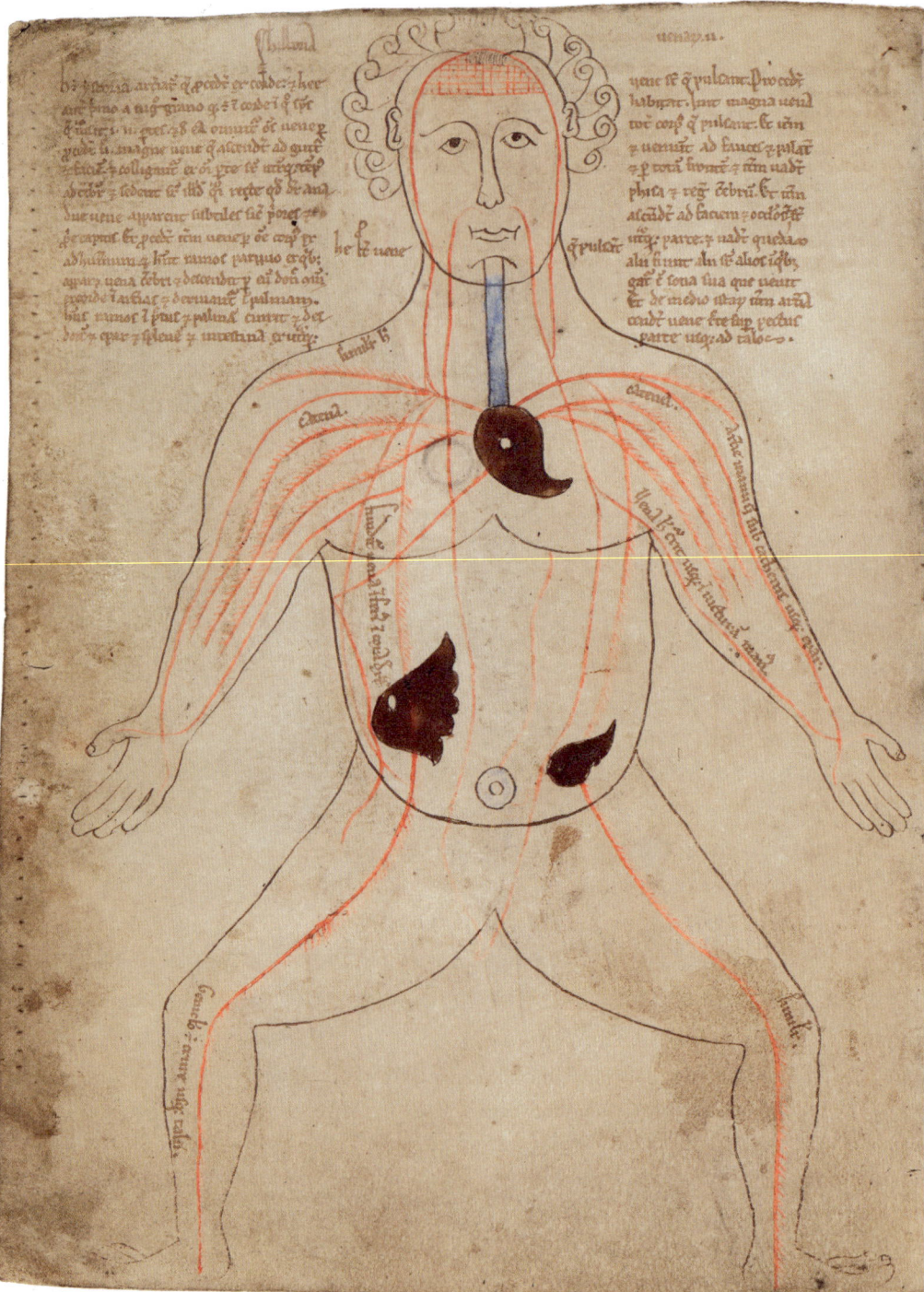

themselves, and some of the text encroaches on the images, suggesting that it may have been added later. Moreover, much of the text is concerned with the production of semen and with theories in respect to reproduction. A more directly relevant part has been translated as follows:

> This is the story of the arteries, which process from the heart. And these are the veins [*sic*] which pulsate, for their source is from a black [*sic*] seed in the heart, in which the spirit resides. And from there processes a large vein, which divides into two parts, the right and the left, and from that very vein proceed throughout the body all the veins which pulsate.[6]

The significance of the black dot or seed (*nigrum granum*, the 'black grain') inside the heart is debated. It is at once both the origin point of the arteries and the seat of the *spiritus* ('spirit'). The text reads: 'Procedunt autem a nigro grano quod est in corde in quo spiritus habitat' ('Their source is from a black [*sic*] seed in the heart, in which the spirit resides'). The exact definition of the word *spiritus*, both in this specific context and historically, is not precise. It could refer to air or breath, or, more significantly, the entity that we now call the soul (*anima*) or the life spirit (*genius anima*). The Romans also described the *genius anima* as the source of seed in reproduction. The text from which the excerpt above is taken describes the veins (defined as 'vessels that

FIGURE 2.2 (opposite) *Nigrum Granum*, from the Gonville and Caius manuscript 190/223, *c.* 1200, England.

FIGURE 2.3 (overleaf left) From the *Five-Figure Series*, Bodleian Library (Ashmole 399) manuscript, thirteenth century, England.

FIGURE 2.4 (overleaf right) From the *Five-Figure Series*, Bodleian Library (Ashmole 399) manuscript, thirteenth century, England.

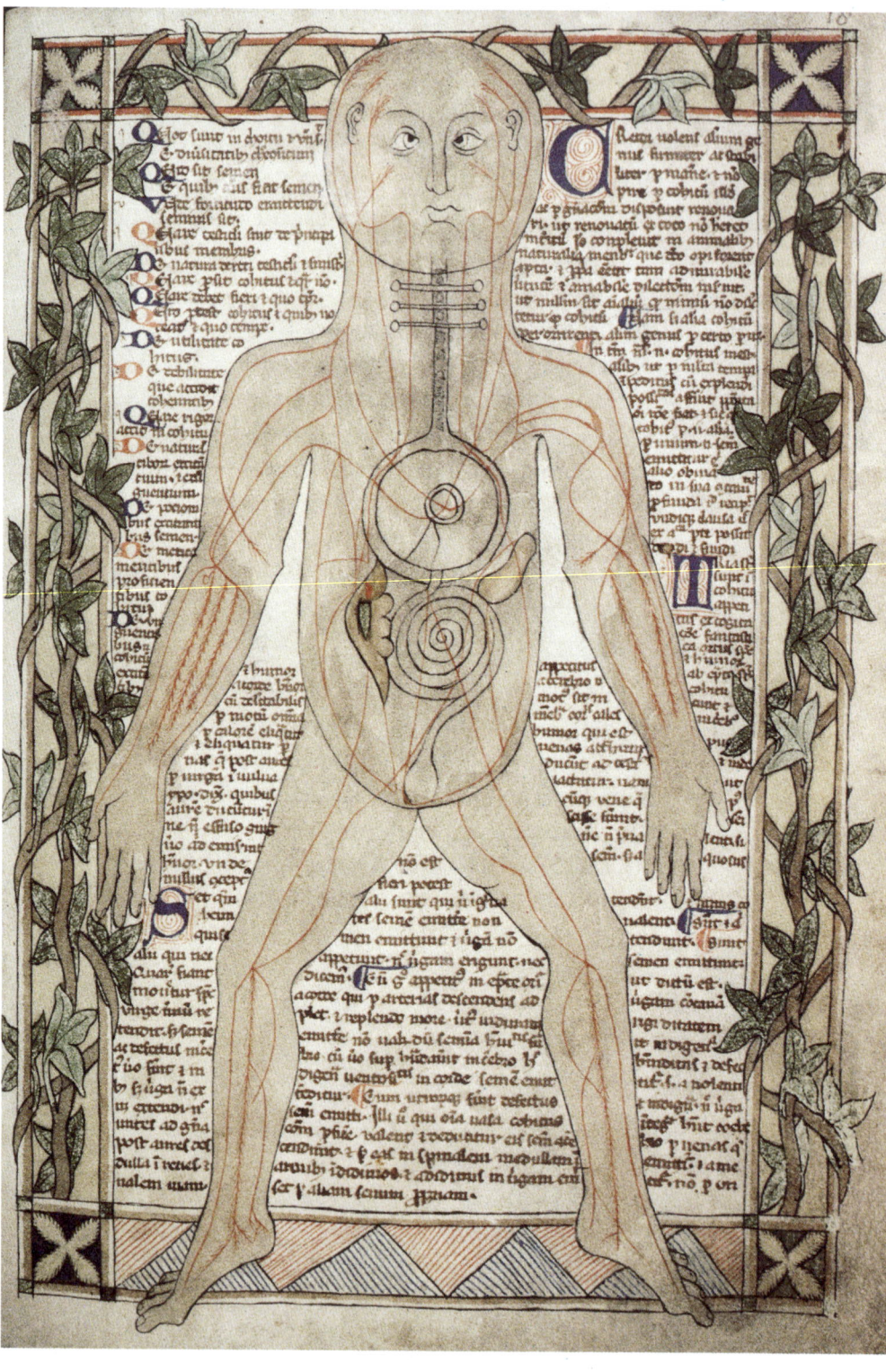

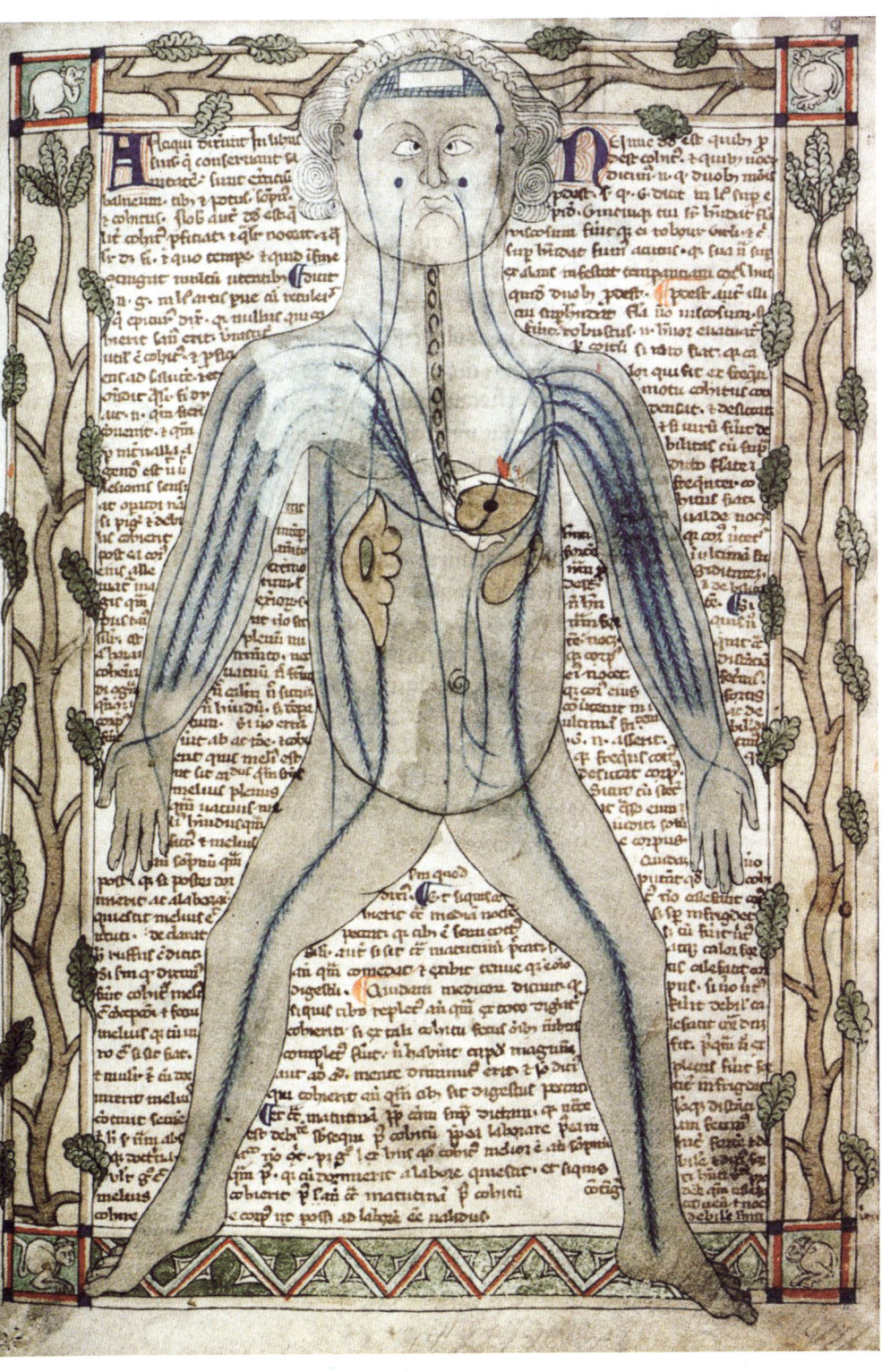

do not pulsate') as originating in the liver and passing up through the diaphragm. On the other hand, the arteries (defined as 'veins that pulsate') originate in the heart, specifically in the *nigrum granum*. We can also see the *anaphusa* (from the Greek for 'network') of arteries on the forehead, also known as the *rete mirabile* ('glorious net'), which protects and nourishes the brain (Figures 2.2 and 2.4).[7] The *nigrum granum* is prominent in the Gonville and Caius (190/223) manuscript (as an uncoloured circle within the brown heart) and appears even more clearly in the Bodleian version (Ashmole 399) (Figures 2.3 and 2.4) as a true black spot. Whatever its purpose, the *nigrum granum* brings back into focus the multiplicity of functions and properties attributed to the heart during this period. The notion of a seed in the heart recalls the ideas discussed earlier, in relation to the Upanishads and to early Judeo-Christian texts.

In his notebooks, Leonardo da Vinci even went so far as to portray the heart in seed-like form, from which the vessels sprouted upwards and downwards, as if the heart were in germination. Clearly, none of this is a coincidence. It seems natural now, as then, that this uniquely dynamic, beating organ must be the seat and the source of life (see Thomas Aquinas, below) and, if so, why not also its means of propagation? We must remember the prevailing urge to unify nature under common laws. If plants grow branching structures from sprouting seeds, then surely the branching vascular structures might have had some similar origin.

The later Persian treatise composed by Mansur ibn Muhammad ibn Ahmad ibn Yusuf ibn Ilyas (*c.* 1380–1422 CE) shows similarities with the Latin versions. Mansur came from a Persian family of scholars and physicians working in the city of Shiraz. In his illustrated treatise, *Mansur's Anatomy* or *Tashrih-i Mansur-i*, the figures are again in the distinctive squatting posture. While the earliest Latin versions date from the twelfth century, the earliest Islamic version is found in the Mansur manuscript held by the US National Library of Medicine and was produced in 1488 (Figure 2.5). Once again, we do not know in what form or by what means the anatomical diagrams of the five systems were made available to Mansur,[8] but it is possible they were received by back-transmission from the works of Western scholars, following their exposure to Arabic scholars' Latin texts.

I should emphasize that, in this period, there was no sophisticated attempt to tie the attributes of the heart to its depiction. Nor would that have been a straightforward task, given the many and varied functions assigned to the organ: it created heat, instigated motion, formed semen and generated spirits to carry out the work of the soul. We might conclude that the 'scientific' treatment of the heart in this period was superficial and, in a sense, impressionistic. It still relied as much on intuited notions as on what could be discovered or described. For the most part, its attributes were mystical. They were summed up perfectly by Dante in *Vita Nuova*[9], possibly with reference to the vital spirits of Aristotle, as 'lo spirito della vita, lo quale dimora nella segretissima camera del cuore' ('the spirit of life which dwells within the most secret chambers of the heart').

Figure 2.6 shows a detail from an English depiction – dating from around the fifteenth century – of a heart within the chest of a pregnant woman (Wellcome MS290). The heart is outlined in brown ink and coloured pink with delicate striation. The left and right sides of the heart are divided by a central bowing line – an exaggeration of the so-called interventricular sulcus (a groove in which the main artery supplying the heart itself runs). The overall effect is to create overlapping tapering ellipses that together make the shape now associated with the heart motif.

Given the preoccupations of the time, it is hardly surprising that the heart was more commonly depicted outside of any medical or anatomical context and became a powerful motif in both secular and religious settings. Common features relating to the many beliefs shared by geographically disparate cultures (and discussed in the previous chapter) soon become apparent. In the Indian, ancient Egyptian and Judeo-Christian traditions,

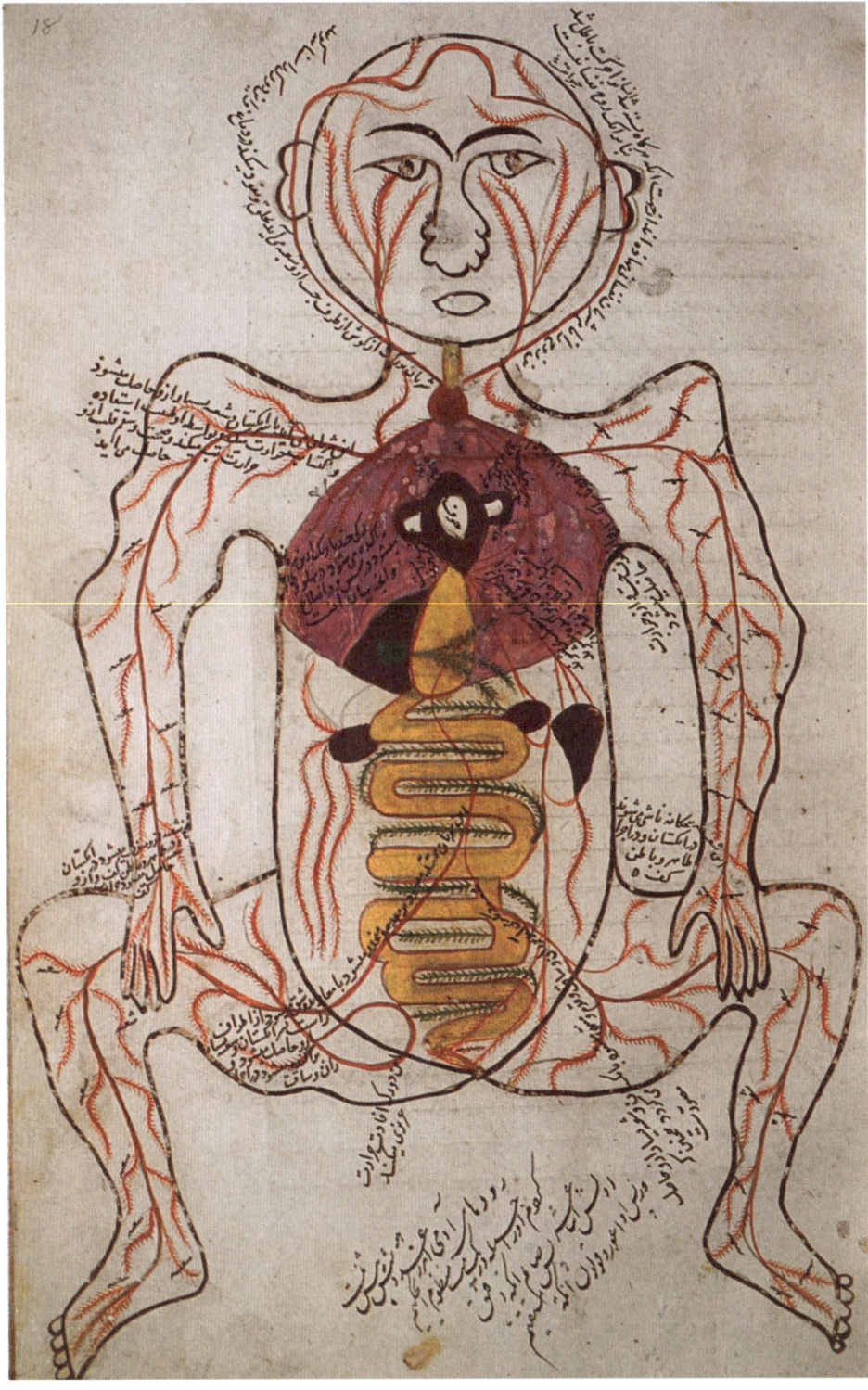

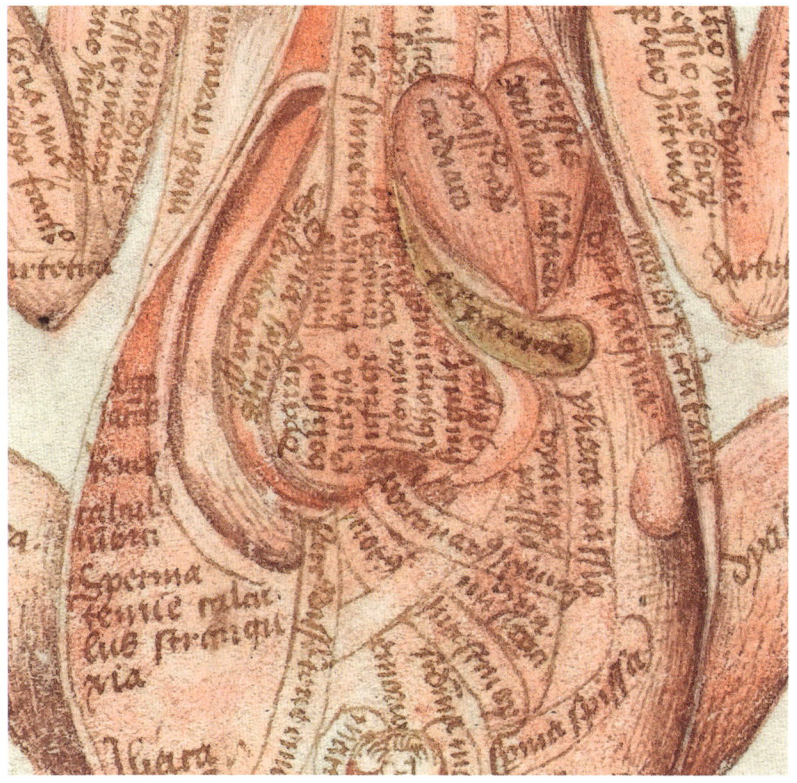

FIGURE 2.5 (opposite) The earliest Islamic version of the anatomical series, *Mansur's Anatomy* or *Tashrih-i Mansur-i*, 1488, Persia.

FIGURE 2.6 (above) Heart within the chest of a pregnant woman (detail), Wellcome MS290, fifteenth century, England.

the heart had been seen as a vessel, a repository of self and a tablet or book. In the Middle Ages, in both the sacred and secular contexts, there was a very firm notion that the heart could carry or contain transferable properties. The third and crucial category for the use of the heart motif, from the medieval period to our own era, is the heart as a seat of love, especially divine and romantic love.

The sacred heart

The heart is represented frequently in Judeo-Christian scripture, where it is mentioned more than a thousand times,[10] often denoting aspects of inner being in relation to understanding, conscience, memory, volition and love. It is often associated with internal writing or inscription – ideas that, given their similarity to the beliefs and practices articulated in the *Book of the Dead*, and to geographical proximity, plausibly have their origins in ancient Egypt.

> Let love and faithfulness never leave you; bind them around your neck, write them on the tablet of your heart. Then you will win favour and a good name in the sight of God and man.
>
> Proverbs 3:1–2 (NIV)

We find further references to the heart in the New Testament – for instance, relating to the Last Judgement and the opening of the secrets of the heart, elaborating on the imagery of reading the contents of the heart by way of a personal revelation.

> You yourselves are our letter, written on our hearts, known and read by everyone. You show that you are a letter from Christ, the result of our ministry, written not with ink but with the Spirit of the living God, not on tablets of stone but on tablets of human hearts.
>
> 2 Corinthians 3:2–3 (NIV)

Indeed, when Gentiles, who do not have the law, do by nature things required by the law, they are a law for themselves, even though they do not have the law. They show that the requirements of the law are written on their hearts, their consciences also bearing witness, and their thoughts sometimes accusing them and at other times even defending them. This will take place on the day when God judges people's secrets through Jesus Christ, as my gospel declares.

Romans 2:14–16 (NIV)

Although the origins of the metaphors of inscribing the tablets of the heart are easily grasped, the florid depictions that follow in painting warrant further exploration. There seems to have been a transition from a notion of dry inscription of tablets to portrayals of the heart that were very much fused with the idea of the heart as an organ. Or perhaps more accurately (hard though it is to imagine), at that time there was no sophisticated concept of the heart as an organ with the distinct physiological roles that we understand today. Instead, natural philosophy, theology, poetry and medicine all employed the same terminology and with overlapping frames of reference. Therefore, to view the depictions with any empathy, we need to dissolve our contemporary delineations of literal science and literary metaphor. It is very difficult, but try to imagine how curious you would be about the workings of your heart if you had no clear understanding of its physiological function.

The Vulgate (common Latin) translation of the Bible was used widely from around 400 CE and became the official Latin Bible of the church after the Council of Trent.* In that translation, the passage from 2 Corinthians (given above in a modern English version) refers to the 'fleshy tablets of the heart' (*tabulis cordis carnalibus*).[11] The origins of the depictions of the heart in the Christian world probably owe much to Augustine of Hippo (St Augustine, *c.* 354–430 CE), the most influential theologian of the early church. A North African Roman, Augustine was born in what is now Algeria and educated in Carthage. His theology

---

* The Council of Trent was an ecumenical council of the Catholic Church held in Trento in Northern Italy (1545–63) in response to the Protestant Reformation.

FIGURE 2.7 St Augustine offers his heart, fifteenth century, Germany.

made much of the moral and spiritual interior life, which for Augustine was firmly centred on the heart as the seat of memory and the record of deeds. In his account of his own conversion, he wrote:

> You rescued my tongue as you had rescued my heart...
> My writing was now done in your service...
> You had shot through our hearts with your charity, and we carried about with us your words like arrows deep in our flesh[12]

In one miniature, a fifteenth-century manuscript copy of Augustine's confessions, a serene Augustine, seated or kneeling, cradles his plump, fleshy heart in one hand as he pens confessions with the other; transcribing, as it were, his heart's inner contents. A much more detailed handling of this subject and many further instances of the heart as a book in medieval texts, including the ultimate extension of the metaphor to books that were formed physically in the shape of the heart motif, can be found in Eric Jager's *The Book of the Heart*.[13]

Besides elaborating on the metaphor of the heart as book, Augustine seems to have propagated much more graphic images, especially in relation to grief and loss, where he referred to his heart being 'broken and wounded and dripping with blood'. In the coloured woodcut version shown in Figure 2.7, Augustine is handing his fleshy, oozing heart to God as a cluster of the faithful gaze upwards, hands clasped in startled awe and adulation.

The eleventh and twelfth centuries saw a wider dissemination of the idea of the piercing of the heart and the beginnings of a cult of devotion to the Sacred Heart. The Cistercian Bernard of Clairvaux, for instance, invoked the image of the pierced heart of Christ, the idea of the heart *as* Christ, and the possibility of communication with the essence of Christ through the heart. This notion was to resurface in a more formalized shape through the work of the nun and mystic St Marguerite-Marie Alacoque several centuries later. Her precise documentation of

FIGURE 2.8 Giotto, 'Charity' from the Scrovegni Chapel, 1306, Padua.

her visions of the wounded heart of Christ were to grow into the influential Cult of the Sacred Heart, originating in France and Italy and propagated by Jesuits from the seventeenth century (to which we will return later).

An early depiction of the heart in a sacred context can be found in the small but exquisite Scrovegni Chapel in Padua. The interior of the chapel is decorated with an uninterrupted fresco depicting the life of Christ in the upper layers and the seven vices and seven virtues along the walls at eye level. The artist was Giotto and work was undertaken around 1306 CE. We are particularly interested in the image of the virtue *Karitas* ('Charity'), which shows a woman in contemporary dress framed in a *trompe l'oeil* stone portal (Figure 2.8). In her right hand is a bowl piled high with fruit and bread, representing her worldly goods. Her left hand is raised, cupping the base of an anatomically quite faithfully depicted heart, the apex of which is within the hands of a representation of God, who leans forward. The woman's gaze seems fixed on the heart. Commentaries on this image seem divided as to whether the woman is giving her heart to God, or receiving the 'good heart' of charity from God. To my mind, the ambiguity of who is giving to whom is less interesting than the fact that the heart is adopted for this purpose, with the implicit idea that the heart is attached to, or contains, some property of charity or kindness.

Throughout the medieval period, mystical manifestations proliferated. Signs and letters were commonly 'inscribed' on the hearts of saints in the Middle Ages. Notable among these were St Ignatius of Antioch (*c.* 110 CE), whose inscription story came to prominence in the Middle Ages, and Chiara Vengente, abbess of the monastery in Montefalco near Perugia.

Botticelli's painting *The Extraction of the Heart of Saint Ignatius* (*c.* 1488), now in the Uffizi, was originally in the church of San Barnaba, a small church in the centre of Florence, just to the north of what is now the railway station and which, after 1356, was affiliated with the Augustinian order in an adjacent convent. The Guild of Doctors and Apothecaries commissioned from Botticelli a grand altarpiece for the church, of which the Guild was the patron. The main altarpiece shows the Madonna enthroned with St Catherine of Alexandria, St Augustine, St Barnabas, St John the Baptist, St Ignatius and St Michael. A scene from the life of each of those saints was depicted in the predella panels leading to the altar.

Ignatius of Antioch was executed under the Roman emperor Trajan for his refusal to worship the pagan gods. When the executioners enquired as to why he repeatedly called on Christ, he replied: 'I have his name written on my heart and for that reason I cannot help remembering it.' A version of the story was recorded in *The Golden Legend*, a collection of hagiographies compiled by Jacopo de Voragine (*c.* 1260), from which we learn that, following his death, Ignatius' heart was excised whole and 'split down the middle, revealing the words "Jesus Christ", in letters of gold'.[14] This account probably marks the start of the dissemination, if not the origin, of the story, and coincided with the highly somatized theologies of the time.

FIGURE 2.9 (overleaf) Sandro Botticelli, *The Extraction of the Heart of Saint Ignatius*, Galleria degli Uffizi, *c.* 1485, Florence.

Botticelli's painting (Figure 2.9) shows the moment of the incision of the heart. Ignatius lies in the foreground, fully robed and mitred on an elaborate cushion. A single longitudinal wound extends from his sternum to his abdomen. His right hand shows the wounds of his crucifixion. Two eager disciples crouch earnestly beside the body, one with scimitar in hand, incising the heart and poised to reveal the golden letters within.

One wonders how Botticelli imagined the heart. His own early training had been as a goldsmith, but presumably he had, at the very least, seen an ox heart or a sheep heart. But what did he imagine was its function or true contents? For certain, he did not see it as a pump. Frustratingly, he offers us few clues, since he chose to depict not the 'reveal' but the moment of anticipation immediately after the incision was made, just before the contents of the heart were disclosed. It was a clever ploy on the artist's part. Botticelli draws the viewer into a scene bristling with expectation, but saves himself from having to commit to a particular version of a fantastical account.

FIGURE 2.10 Santa Chiara de Montefalco, Church of Santa Maria Incoronata, Milan.

Chiara of Montefalco took the somatic internalization of Christ still further. Prior to her death in 1308, she had told the other nuns of the abbey that Christ lived in her heart and nourished her from within. Such was their faith (or, perhaps, their curiosity) that in the days after her death, her sister nuns first removed her heart to a casket and, a few days after that, they incised it to explore its contents. It seems that, on opening the heart, the nuns discovered a crucifix and several other indispensable relics of the Passion, such as the nails that pierced Christ's limbs and the whip of his flagellation. Crucially, all were fashioned from the flesh of the abbess's heart. News of these claims spread and, not surprisingly, were carefully scrutinized. Although the findings were initially challenged, they were eventually 'verified' by the local bishop and St Chiara was canonized, in the nineteenth century, by Pope Leo XIII.[15] The depiction of St Chiara in the Church of the Coronation in Milan (Figure 2.10) is typical: it shows her standing in life and holding her heart in front of her. Her heart has the familiar heart motif shape and bears within it a miniature Christ on the Cross. Less sympathetic depictions show Christ seemingly delivering a full-size crucifix through the chest wall of the bewildered abbess. Arguably, such literal depiction misses something of the miracle of Christ residing within the heart of the devoted nun. Sometimes renderings appear more analytical, even forensic, such as woodcut prints in which Chiara's isolated heart contains the critical paraphernalia carefully laid out and itemized as exhibits, perhaps for inspection by the church.

Catherine Benincasa (St Catherine of Siena; born in Siena *c.* 1347, died in 1380, canonized in 1461) drew the heart deeper into Christian theological thought and used notions of the heart in metaphors for political effect. Deviating from the Pauline notion of the church as 'one body in Christ', with Christ as its head, she wrote in her letters of Christ as the heart – at the centre: 'In the middle of the vineyard [the church], He placed the vessel of his heart, full of blood to water the plants within it so they don't dry out.'[16]

This was a significant distinction: the move from head to heart implied a shift from 'command and control' to central support and nourishment. Indeed, Catherine expanded her theology to show her support for the centralized political authority of Pope Urban VI, whom she portrayed as Christ's earthly representative and at the heart of the body of the church.[27] In emphasizing the central nutritive elements, she sought to diminish the perceived threat of top-down papal control. Her structural organization of the church aligned with Aristotle's views of the heart as the centre, the home of the soul and the source and essence of life. Accordingly, in her recollection of her mystical experiences, Catherine also placed the heart centrally, recording in a letter to Pope Urban that she had provided God with her heart, effectively to nourish the church in a time of hardship, and that God had taken out her heart and pressed it, squeezed out her blood, to nourish the arid church. In other words, in her version, Catherine was the giver and the prime mover. In her biography, written by her confessor, Raymond of Capua, the story is given a different slant; Catherine is reduced to a more passive role:

> Later on one occasion, Catherine prayed fervently, like the Prophet. The Lord came to her in a vision, and he opened her left side… and removed the heart, so that she remained without a heart inside… Another day… she was surrounded by a light from the sky, and in the light appeared the Lord, holding in his sacred hands a human heart, red and shining*… The Lord opened her left side again. He put his own heart, which he held in his hands, inside her, saying, 'Here dearest daughter since I took away your heart before, now I bequeath to you my heart, so that you may live forever.'[17]

The story is depicted in a painting by Giovanni di Paolo in the Metropolitan Museum, New York, who seems to have adopted Raymond's version and combined these miraculous events into a single image (Figure 2.11). If we look carefully at the heart itself, this fifteenth-century depiction has again assumed the heart motif shape, and it has

FIGURE 2.11 Giovanni di Paolo di Grazia(1403–82), *Saint Catherine of Siena Exchanging Her Heart with Christ*, Siena.

additional features that must have been quite deliberate. Firstly, as in the image of St Augustine (Figure 2.7), the heart is fleshy and dripping with blood. Secondly, both the heart and the blood are the same rich red, contrasting with the subdued tones of the rest of the image, and of St Catherine in particular. In the process of the exchange of hearts – a prototypic transplantation – something both fundamental and substantial was being transferred.

The extraordinary accounts of St Ignatius, Chiara of Montefalco and St Catherine were well known, but not isolated, stories of the heart-inspired mysticism of the period. From our twenty-first-century vantage point, it is very hard to approach the mindset of an age that holds such stories to be plausible. But that misses the point of the core question, which historian Jack Hartnell frames well:

> Of all the places where Christ could dwell within Santa Chiara de Montefalco's body, why did it make sense to the sisters of Santa Croce, indeed to the saint herself, that he might have miraculously materialised within her heart? What was it about the organ that enticed sister Francesca to slice deeper into its chambers? And what else could this central vital body part prompt in the minds of mediaeval people?[18]

The secular heart

We hardly need to rehearse how our language, idioms and meta-phors have become laden with references to the heart ('heartbreak', 'sweetheart', 'heartfelt', etc.). These allusions are usually, though not exclusively, directed towards deep emotion, often associated with love. From around the fourteenth century onwards, very literal, graphic rep-resentations began to appear in secular depictions of love and the heart.

In his book *Love, A History*, Professor Simon May proposes an evolution in the history of Western love over four transformations that incorporate both sacred and secular variants. If we examine the properties attributed to the heart and the way its image is used, we find that depictions of the heart track closely with the phases of cultural development that May brings forward:

> The first transformation concerns the *value* of love. Between Deuteronomy and Augustine – so for well over a thousand years, until the mid-fifth century CE – love is gradually made the supreme virtue. Hebrew scripture commands that God be loved with 'all your heart and with all your soul and with all your might'. ['Love the Lord your God with all your heart and with all your soul and with all your strength'; Deuteronomy 6:5 (NIV)]. Jesus elevates love of God and neighbour to the most important biblical commandments; John the Evangelist says that ultimate reality – God – is love; Paul and then St Augustine see love as the root of all true virtue.
>
> In the second transformation spanning the fourth to the sixteenth centuries CE, from Augustine to Bernard of Clairvaux to Thomas Aquinas, and beyond him to Luther, humans are given unprecedented – literally divine – *power* to love. By developing the idea of love as a gift of divine Grace, human beings can at least in principle become divine through love and even achieve friendship with God...
>
> The third transformation emerging in the eleventh century and culminating in the eighteenth concerns the *object* of love.
>
> ...This prepares the ground for a fourth transformation – beginning in the eighteenth century with Rousseau, and still very much underway – which concerns *the lover*, who becomes authentic through love.[19]

We have already seen how some elements of the first and second transformations have been manifested in depictions of the heart. There are ample examples of words and deeds recorded in the heart and disclosable to God. The accounts of St Catherine of Siena exemplify the heart as an organ of power in the medieval Christian church, with the possibility of direct communication with Christ, and even transfer of certain properties by the exchange of hearts.

By the twelfth century there was also an emerging literature of love in the secular context. Its origins owed more to the classical gods of love (Eros and Cupid) than to the Judeo-Christian tradition of God-given love. Classical depictions of Eros were typically of a winged, athletic adult male, often with a bow and arrow, and sometimes a sleeping child. Cupid, on the other hand, was usually shown as a young boy playing, driving a hoop, chasing bees or butterflies or flirting with nymphs. Sometimes, his mother (Venus) would be depicted scolding him over some mischief or other. He was also sometimes shown with a bow or throwing darts to wound the lover, giving physical form to the classical notion of the pain of love, as described, for example, in the poetry of Ovid, writing in the first century BCE and first century CE.

From the twelfth century, the notion of *amour courtois* ('courtly love'), emerged in the royal and ducal courts of Aquitaine, Provence, Champagne and Burgundy. According to this tradition, the lover idolized his unattainable mistress. She would typically be married and of high social rank; the knight suitor would be unworthy of her attention. He would demonstrate his faithfulness and obedience through noble acts, bravery and by suffering various ordeals that would prove his ardour and devotion. Typically, the knight's love was unrequited. Courtly love was a prominent feature of the work of the troubadours – lyric poets and musicians working from the late eleventh century in what is now southern France (Occitania) – who were responsible for spreading the idea and its conventions across Europe. In this literary context, the heart was a readily intuited motif for the pains and tribulations of the courtly lover.

Indeed, excruciating wounds of the heart are precisely what German printmaker Casper von Regensburg portrayed in his allegorical painting of *c.* 1485 (Figure 2.12). In the centre of the image, Frau Minne, a personification and object of courtly love, stands alluringly in a state of undress (except for a wispy shawl draped around her hips), while around her an array of hearts are subjected to a variety of mechanical abuses and torture: by piercing, cracking, squashing, sawing, burning and crushing underfoot.* One of the hearts, on which she stands, is punctured by her right foot. This is a brutal scene. Meanwhile, the physically diminished suitor kneels at her feet, wearing a bloodied smock, begging relief from the pain inflicted by the unattainable object of his affections. In other representations of the period, such as one found in Zurich, Frau Minne is shown removing the beating hearts from her suitors' breasts, and in a gentler version painted on the underside of the lid of a coffret (a chest) and now in the Metropolitan Museum of Art, the wounded heart, pierced by Frau Minne's arrows, is presented to her calmly by the suitor, as he leans somewhat cautiously away (Figure 2.13). For us, the significance of these images is severalfold. They speak directly to a concept of the object of human love, and they provide an early example of the secular use of the heart motif and, most importantly, its adoption as a shorthand for the heart as a sentient organ for the articulation of painfully intense emotions surrounding love and desire.

Intriguingly, the concept of the weight of the heart to gauge the reciprocation (or otherwise) of love was also used in this period. In a series of images from a fourteenth-century tapestry, Frau Minne is presented with a winged heart, which she moves to puncture with an arrow.[20] In the same series, tucked away in the bottom right of six

---

* Frau Minne appears in the era of Middle High German literature (twelfth to fourteenth centuries), during which a form of lyric song known as *Minnesang* ('love song') flourished.

circular panels, the lady holds a set of weighing scales. Each of the two small slings at the end of the balancing arm cups the apex of a heart (in heart motif form). To the consternation of the suitor, the balance tips towards him as the lady points and smiles, perhaps indicating the imbalance in devotion or worthiness between their two hearts, or possibly alluding to the Biblical prophecy of Ezekiel, in which a heart of stone is replaced with a heart of flesh: 'I will give you a new heart and put a new spirit in you; I will remove from you your heart of stone and give you a heart of flesh' (Ezekiel 36:26).

The notion of the explanted heart as a distillate of self could also serve practical purposes. If some notable person died away from home, it might have been desirable – for example, for political or devotional reasons – to repatriate their body for burial. Clearly, though, strapping a rapidly bloating, decomposing corpse to a horse-drawn cart for transportation would have been problematic. Instead, the removed, embalmed heart was a more sanitary and deliverable option.

In this vein, Dervorguilla of Galloway (*c.* 1210–90 CE) offered a devotion after the death of her husband, John de Balliol. Dervorguilla built Sweetheart Abbey, in Galloway, as a resting place for her and for his heart. The Scottish poet Andrew of Wyntoun (*c.* 1350–1425 CE) records its origin:

> When the Balliol that was her lord,
> That spousit her, as they record,
> Had sent his soul to his Creator,
> Ere he was laid in sepulture,
> His body she gert open tite
> Angert his hart to be tane out quite.
> And that ilk hart, as men said,
> She balmed ot and gert be laid
> Into a coffin of evore
> The which she gert me made therefor,

Enamellit, partfitly dicht
Lokkit and bunden with silver bricht
…
She ordanit in her testament
And gave then bidding verrament
That his hart they should be ta'
And lay between her pappis twa,
When they suld make he sepulture
To her lord she did this honour…
She founddit then in Galloway,
Of Cistew's orer, ane abbay,
Dulce cor she gert them all.
(That is Sweetheart) the abbay call.[21]

If we have begun to address the question of *how* the heart came to its cultural prominence, we have hardly considered *why* this organ, above all others, came to assume and retain this position.

It is hard to overstate the depth of fascination that must have built up around this organ, especially before its physiological function was understood. Here is an organ that moves ceaselessly through life and stops at the moment of its extinction. Little wonder that Aristotle focused so much of his attention on the heart and imbued it with such fundamental properties. Many subsequent thinkers and scholars were similarly transfixed. Prominent among them was the priest and theologian Thomas Aquinas.

Aquinas (1225–74 CE) was active at a critical juncture of Western culture, at a time when the works of Aristotle had become available in Latin translation, from the less accessible Arabic and the original Greek. After his early studies at Montecassino, he moved on to the University

FIGURE 2.12 (overleaf) Casper von Regensburg, *Frau Minne and the Suitor*, *c.* 1485, Germany.

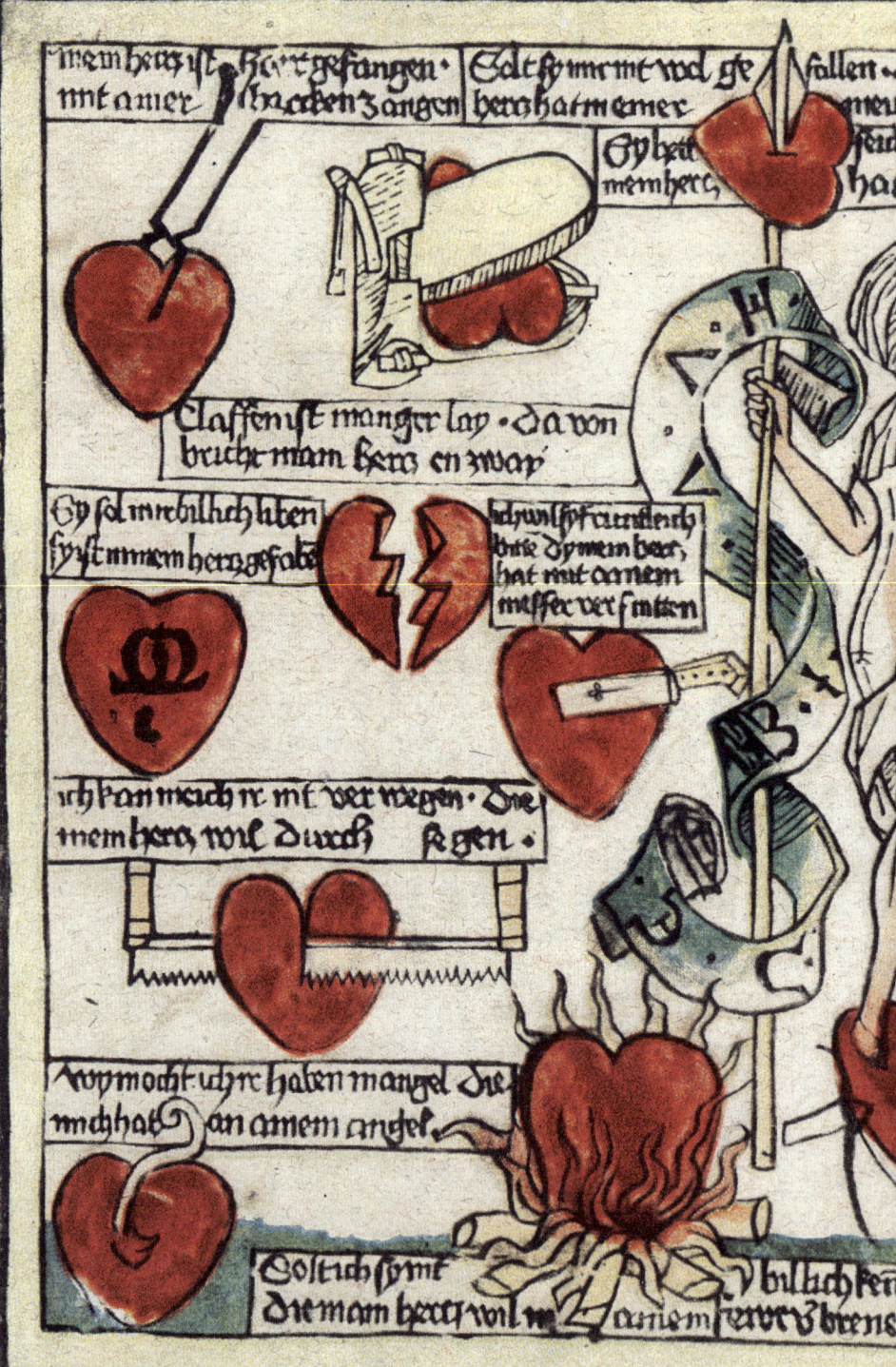

mein hertz ist · hertt gefangen ·
mit a mer   in roten zangen ·

Solt sy me in wol ge
hertz hat me iner

Ey het
mein hertz

fallen ·
hat

Claffen ist manger lay · da von
bricht man hertz en ain

Ey sol in rebilich liben
sy ist in mein hertz gefalt

Ich wil sy frauntlich
bitte dy mein heer
hat mit ainem
messer verschnitten

ich kan mich re mt ver wegen · Die
mein hertz woll dur ach   se gen ·

wy mocht ich re haben mangel die
mich hat   an ainem angel ·

Solt ich sy mt
die man hertz wil m

bilich seit
ainem feuer verbrens

ch hab nit wol genossen Dy mein hat durch stossen   wy solt ich ir vergessen mein hertz hat in der press

Dy kan mich wol vnter weisen · Dy mein hertz vecht in ainer rosen ·

wy nocht ich ir ... hat mein hertz in ... sie fast

Sigt mir frend vnd trost · Dy mein hertz hat uff ainem rost ·

Der beser mant · hat mein hertz ser verwunt

Schen hubsch vnd auch mich ... Erloss der pein vnd in die arm dein.

O fraulein hubsch vnd auch mich ...

Ich wil ir stette trou loben · Dy mein hertz hat in meinem cloben

calper

FIGURE 2.13 Frau Minne and the suitor's heart, *c.* 1325–50, Germany.

of Naples, where he met members of the new Dominican Order,* whom he joined before heading north to study with Albertus Magnus, author of a commentary on the Aristotelian corpus. He completed his studies at the University of Paris.[22] Aquinas emerged from his studies with a decidedly Aristotelian outlook, and he was instrumental in reconciling the pagan writing of Aristotle with the teachings of the Christian church. In his letter *De Motu Cordis* (*On the Motion of the Heart*), he addressed the role of the heart. The precise identity of the recipient of Aquinas's letter ('Master Philip') is uncertain, but he seems to have approached Aquinas as an authority and asked about the significance of the heart. For his reply, Aquinas drew from the works of Aristotle,[23] including his biological text *On the Motion of Animals*. For both scholars and philosophers, motion was key to addressing the heart's importance in the body and the relationship between the heart and the soul.

I am certainly not the first to observe that 'scientific' endeavour, or just more general human attempts at understanding, are hugely influenced by the prevailing ways of seeing the wider world at the time. That might relate to the conceptual: for instance, we might interpret our observed world from a religious standpoint or through a rigorously scientific prism. Or our view might be influenced by new technology: as, for instance, when the perception of the natural world was transformed by the microscope and the telescope in the seventeenth century. For Aquinas, two particular external references would have been prominent in his contemporary world. One was consideration of the source of body heat. With no comprehension of cellular metabolism and energetics or exothermic chemical reactions, the moving heart – filled with warm blood in life but cooling after death – must have been enthralling. From Aristotle (whom Aquinas called 'the Philosopher') onwards, the heart had been

---

* A Catholic mendicant order founded and named after Spanish preacher Dominic de Guzman. Founded in 1215 to preach the gospel and oppose heresies, it was particularly known for having an intellectual approach that appealed to philosophers and theologians.

seen as a kind of internal furnace. Aquinas wrote in his *De Motu Cordis*:

> Some others say that the principle of this motion in the animal is heat, which being generated by spirit moves the heart. But this is unreasonable. For the deeper principle is more likely to be the primary cause. But the motion of the heart is a deeper principle in the animal and more contemporaneous with life than even warmth. Therefore, warmth is not the cause of the heart's motion, but on the contrary the heart's motion is the cause of warmth. Thus, the motion of the heart is a natural result of the soul, the form of the living body and principally of the heart.[24]

The second consideration related the movement of the heart to the movements of the planets:

> Perhaps this is why some who have understood this go on to say that the heart's movement is caused by an intelligence, for they think that the soul comes from an intelligence (which is similar to what the Philosopher says… about the movement of heavy and light things coming from a generator that gives the form which is the principle of their motion). But it is important to note that every property and movement is a result of a form in a particular condition. So as a result of the form of a subtle element like fire, there is motion to a subtle place, namely upwards motion. Now the most subtle form on earth is the soul, which is most like the principle of the motion of the heavens. Thus, the motion that results from the soul is most like the motion of the heavens. In other words, the heart moves in the animal as the heavenly bodies move in the cosmos.
> We should note that there is a difference between the principle of the heavenly motion and the soul. The former is not moved in any way at all, neither essentially nor incidentally, but the

sensitive soul, although unmoved essentially, is moved inciden-
tally. Thus, different types of sensations and emotions arise in
it. So, whereas the heavenly movement is always uniform, the
heart's movement varies according to the different emotions
and sensations of the soul. For the sensations of the soul are not
caused by changes in the heart, but just the opposite is the case.
This is why in the passions of the soul, such as anger, there is
a formal part that pertains to a feeling, which in this example
would be the desire for vengeance. And there is a material part
that pertains to the heart's motion, which in the example would
be the blood enkindled around the heart.
Now although some change occurs in the heart's motion because
of different sensations and feelings, nevertheless such change is
involuntary, for it does not come about through the command
of the will.

It seems to me that in these passages, Aquinas crystallizes precisely
why the heart has assumed its primacy. Remarkably, he did so 250
years before Copernicus proposed his heliocentric theories and 350 years
before Harvey's elucidation of the circulation of blood and illumination
of the heart as a pump.

I want to align with Aquinas's thinking and draw out what I believe
are the three key elements. Firstly, that the heart is a *prime mover*, akin
to the autonomous and ceaseless movement of the planets. Secondly,
that it beats with no *conscious* input, but, thirdly, that it is *responsive*, as
moved by the soul. In other words, the heart exists within us but apart
from us, as a transducer of emotions and passions beyond our direct
conscious control. In my view, this distillation by Aquinas explains
almost completely our enduring attachment to the heart. Later we
will see how, despite an expanding understanding of how the heart
and circulation function, and even of the basis for its automaticity and
responsiveness, these primal intuitions have persisted, undiminished.

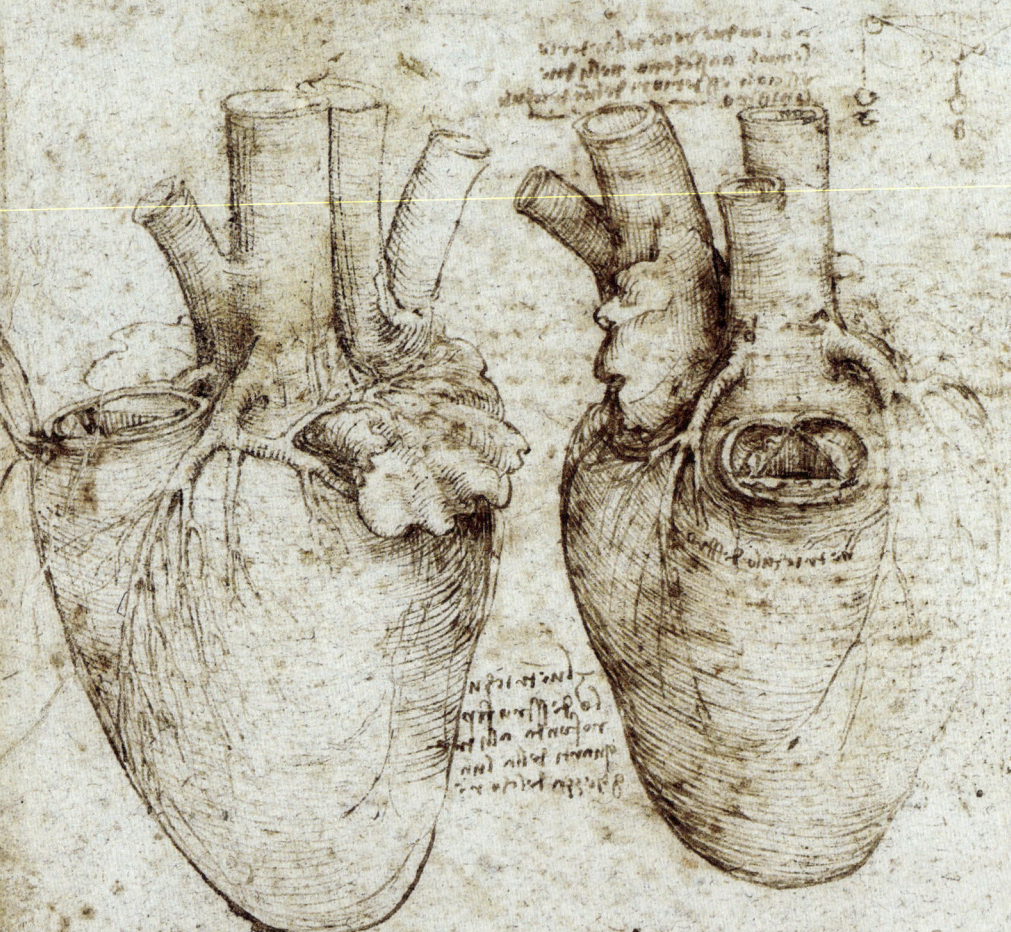

Opening the Gates: Leonardo and the Heart

In tracing the great revolution of learning, which happened in the fifteenth century, I am enabled to carry the history of the improvement of Anatomy farther back than has been generally done by other writers; and to introduce into the annals of our art, a genius of the first rate, Leonardo da Vinci, who has been overlooked because he was of another profession and because he published nothing upon the subject. I believe he was, by far, the best Anatomist and physiologist of his time…[1]

WILLIAM HUNTER

1794

P lacing Leonardo da Vinci into this broadly chronological narrative is highly problematic. Firstly, he was essentially anachronistic, both in the rapidity of the evolution of his thinking and in the sophistication of his ideas. His studies in anatomy, best illustrated by his work on the heart, bridged remote intellectual landscapes. In his 1916 attempt at a psychoanalytical study of Leonardo, Sigmund Freud made the point somewhat more eloquently: 'According to Merezhkovski's beautiful comparison he was like a man who awoke too early in the darkness, while the others were all still asleep.'*[2] Secondly, his work was not published in his lifetime and was not widely known. It took other thinkers and investigators in this field almost a hundred years to begin to catch up with him. So, I will attempt to present Leonardo's main works on the heart, and the context in which they were undertaken, in such a way that the reader can form an opinion on their placement and the extent to which they proved a new gearing or pivot point – albeit one that was out of step with the rest of the machine.

Leonardo's contributions to this story are both profound and strangely peripheral. As we shall see, he contributed a detailed, highly analytical understanding of the workings of parts of the heart (especially the valves) that far exceeded that of his predecessors and peers. And yet, at the same time, he never truly grasped the place of the heart as a pump in a circulatory system. His work provided an intellectual ratchet between the unquestioning acceptance of the inherited wisdom of the greats (Hippocrates, Aristotle, Galen) and the emergence of a new critical science of personal observation (of which he was the unquestionable master) and experimentation that characterized Renaissance enquiry. Even as I consciously resist subliminal acceptance of Leonardo as the unquestioned master, I do not believe that it is an exaggeration to state

---

* Freud was referring to Dmitry Merezhkovsky (1865–1941), a Russian novelist, poet, religious thinker and author of *Leonardo da Vinci: The Resurrection of the Gods* and *The Romance of Leonardo da Vinci*.

that his brilliant, single-handed endeavours over some twenty-five years eclipsed the cumulative work of all others in the field for at least a century. But, remarkable to relate, Leonardo's studies were not published and, as a result, were not adopted into the 'scientific' mainstream during his lifetime; they instead remained an obscure, if dazzling, sideshow.

Leonardo established his enduring reputation as a painter in his lifetime. Less well known is the extent to which, in the latter half of his life (from around 1490), he was dedicated to the observation, exploration and depiction of nature. In Leonardo's teleological view, humans and animals existed within an ordered world. He strove to understand human form and function in a context of universal laws of nature, in which all elements existed by purposeful design.

The scope and depth of his studies in anatomy, comparative anatomy, foetal development and what amounts to physiology were immense and are documented in heavily annotated diagrams on large format sheets that were subsequently compiled and bound. His work reveals an extraordinary evolution in his philosophy of enquiry, methodology and understanding, which was all the greater since he was starting not with a blank sheet of paper but with a substantial legacy of entrenched misinformation that had descended from the classical scholars and was often endorsed by the church.

Leonardo's anatomical work starts with the representation of long-established classical doctrine but moves remarkably swiftly to groundbreaking first-hand observations. As such, his approach does not merely reflect personal adjustment or development; it anticipates how the extended 'scientific' community would begin to operate from the seventeenth century onwards, and it reflects the broader spirit of his place and time.

Jacob Burkhardt's seminal analysis (*Die Kultur der Renaissance in Italien*, 1860; *The Civilization of the Renaissance in Italy*) of the position of the Renaissance in relation to the Middle Ages has become somewhat unfashionable in certain quarters today, but it speaks with utter precision to the ossified state of anatomy that Leonardo confronted and single-handedly overcame:

> In the Middle Ages both sides of human consciousness – that which faces the outside world and that which is turned towards man's inner life – lay dreaming or only half-awake, as if they were covered by a common veil, a veil woven of faith, delusion and childish dependence. Seen through this veil reality and history appeared in the strangest colours, while man was only aware of himself in universal categories such as race, nation, party, guild or family. It was in Italy that this veil was first blown away and that there awoke an objective attitude towards the state and towards all the things of this world while, on the other side, subjectivity emerged with full force so that man became a true individual mind and recognised himself as such.[3]

For all the charm of the earlier explorations and depictions, they bore little relation to any anatomical reality. Leonardo approached a range of questions with an utterly new mindset – one that reflected the prevailing culture of individual endeavour and bold enquiry.

Moreover, when we look at his exquisitely drawn anatomical depictions, it is all too easy to forget that the human cadaver does not simply offer up its tissues in the crisply delineated form that Leonardo portrays them. As anyone who has undertaken morbid dissection will agree, even in the best of circumstances, the tissues resist exploration. They adhere to each other, matted together with sinews and membranes and coated with friable yellow fat that fragments, leaving sticky globules strewn along the more interesting structures. On deeper exploration, the tissues are easily damaged: nerves and vessels can be severed, cavities punctured, fibres divided. Factor in the reality of dissection in the fifteenth century, without refrigeration or chemical fixation of the tissues, and the whole undertaking must have been highly technically and physically demanding. Yet Leonardo not only described the morphology and relations of the organs and their component parts, he hypothesized their function and, on occasion, devised ingenious experimental systems to test his conjectures.

In the current age of virtually instant communication of data, where images can be acquired and transmitted in seconds, it is hard to conceive that the painstaking anatomical work of Leonardo went unpublished and largely unrecognized for almost 400 years. Though his endeavours in anatomy were praised in Vasari's biography (*The Lives of the Most Excellent Painters, Sculptors, and Architects*, 1550), and despite its virtuosity, Leonardo's prolific *oeuvre* was not published until the nineteenth century.

Vasari wrote:

Leonardo then applied himself, even more assiduously, to the study of human anatomy, in which he collaborated with that excellent philosopher Marc'Antonio della Torre… one of the first to illustrate the problems of medicine by the teachings of Galen and to throw true light on anatomy, which up to then has been obscured by the shadows of ignorance.

…in this he was wonderfully served by the intelligence, work, and hand of Leonardo, who composed a book annotated in pen and ink, in which he did meticulous drawings in red chalk of bodies he had flayed himself.[4]

The sixteenth-century historian Paolo Giovio[5] records Leonardo's intention to publish the copperplate engravings of his works, though he never did so. It may have been that the collaboration with Marc'Antonio della Torre, who was professor of anatomy at the Universities of Pavia and Padua, was cut short by the latter's death from plague in 1511. Had Leonardo published, he would have anticipated Andreas Vesalius's highly disruptive and influential *De humani corporis fabrica* (*On the Structure of the Human Body*, 1543), which probably would have fundamentally changed our perception of Leonardo as a 'scientist'.

In fact, he bequeathed the unpublished notebooks to his pupil and companion Francesco Melzi, but the notes were exactly that: a large volume of unbound papers.[6] From the Melzi family, the papers were sold on to the sculptor Pompeo Leoni, who mounted the drawings into several large albums.[7] Following Leoni's death in Madrid in 1608, the collection was broken up in a series of sales in Madrid and Milan.[8] By 1630, a sizeable part was in England, owned by Thomas Howard, 2nd Earl of Arundel, and by 1690, the works were in London, in the possession of William III and Mary II. Exactly how they entered the Royal Collection is unknown; they were probably acquired, along with many other Renaissance drawings, by Charles II.[9] The drawings were stored with drawings by Hans Holbein and remained hidden until the reign of George III in the mid-eighteenth century.

William Hunter (1718–83), Scottish anatomist and obstetrician and professor of anatomy at the Royal Academy of Arts, gives a striking account of his shock and admiration on the 'discovery' of Leonardo's anatomical drawings. His lectures, published posthumously in 1784, tell the story:

> In tracing the great revolution of learning, which happened in the fifteenth century, I am enabled to carry the history of the improvement of Anatomy farther back than has been generally done by other writers; and to introduce into the annals of our art, a genius of the first rate, Leonardo da Vinci, who has been overlooked because he was of another profession and because he published nothing upon the subject. I believe he was, by far, the best Anatomist and physiologist of his time.[10]

Hunter refers to Vasari's account, recording the safekeeping of Leonardo's notes by Melzi together with a portrait of Leonardo, and how:

those very drawings are happily found to be preserved in his Majesty's great collection of original drawings. Mr Dalton, the King's librarian, informed me of this and at my request procured me the honour of leave to examine them. I expected to see little more than such designs in Anatomy as might be useful to a painter in his own profession. But I saw, indeed with astonishment, that Leonardo had been a general and deep student. When I consider what pains he has taken upon every part of the body, the superiority of his universal genius, his particular excellence in mechanics and hydraulics and the attention with which such a man would examine and see the objects which he was to draw, I am fully persuaded that Leonardo was the best Anatomist at that time in the world.

The notebooks were finally published in a series between 1898 and 1916, almost exactly 400 years after Leonardo completed them. In 1980, Kenneth Keele and Carlo Pedretti published a complete facsimile of the anatomical studies (with photographs obtained by constructing a custom studio at Windsor Castle), along with a translation of Leonardo's script and an erudite commentary.* [11]

To begin to appreciate the challenge facing Leonardo, and the originality and magnitude of his undertaking, we must first recognize that at the beginning of the sixteenth century the most important anatomical works were purely *textual* descriptions. Where they existed at all, prevailing representations of the heart were schematic and basic, but images and diagrams were not common, reflecting both the ignorance of anatomical detail and the technical difficulties in their reproduction.

---

* My references to Leonardo's annotations of his drawings rely heavily on studying these giant volumes, the *Corpus of Anatomical Studies*, that the staff of the Bodleian kindly extracted from the stacks for me on multiple occasions.

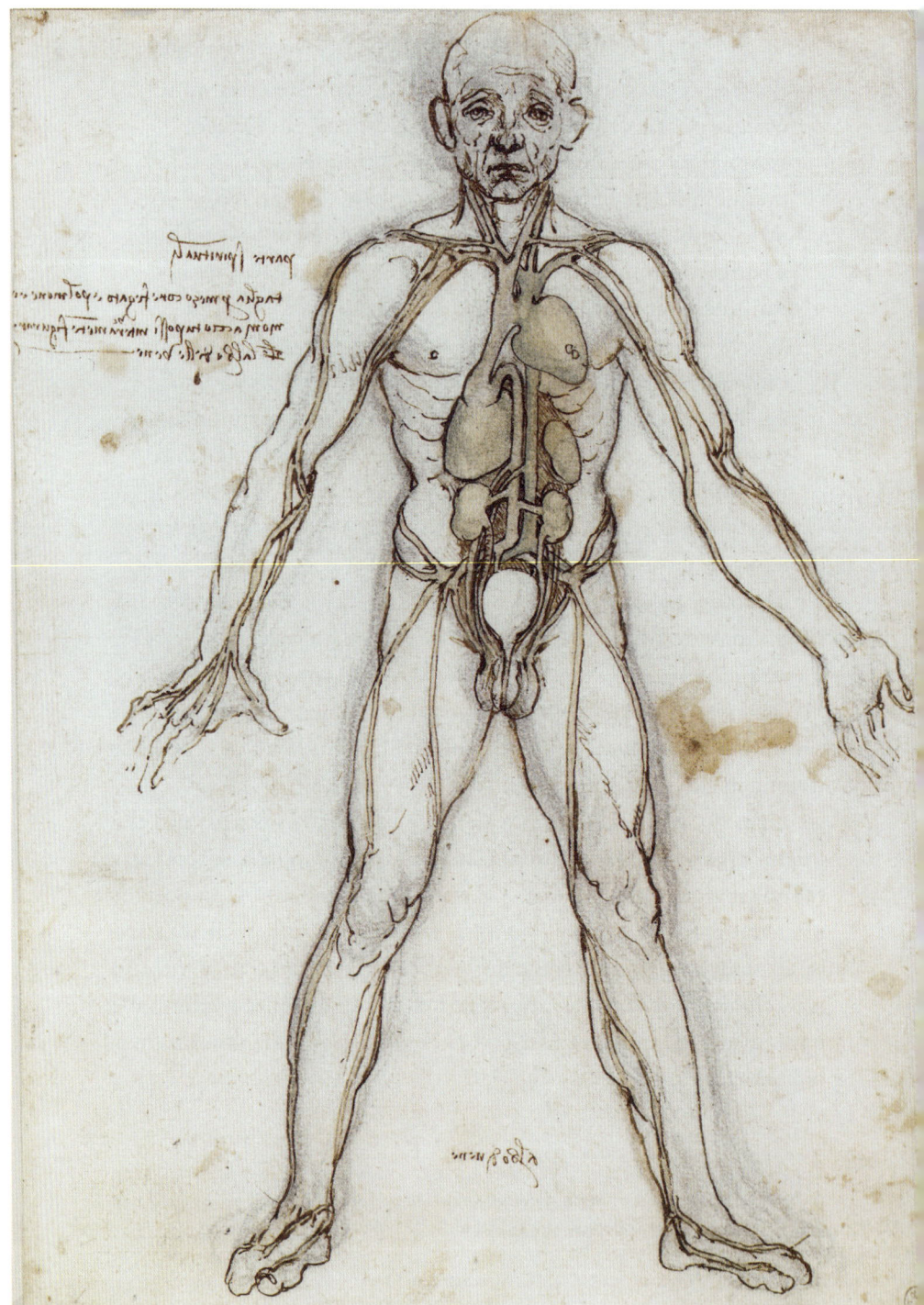

At the start of his endeavours in anatomy, in common with other artists of the time, Leonardo's early drawings focused on visual compilations of the textual descriptions of established authorities (Plato, Aristotle, Hippocrates, Mondino de Luzzi, Galen and Avicenna, whom he cites by name).[12]

Martin Clayton of the Royal Collection Trust at Windsor Castle, a Leonardo scholar, writes:

> Before 1510 Leonardo's method had been to interpret what he saw in the light of what he knew (or what he thought he knew), recording this interpretation; the drawing was the end product of his reasoning. After that date, the drawing usually came first and was the basis for his investigations of the functions of observed form. Leonardo's greatest skills were as an observer and a recorder and thus when his drawings were no longer limited by his imperfect knowledge, he produced some of the most lucid and accurate anatomical illustrations in the history of science.[13]

In fact, for current purposes, Leonardo's anatomical drawings can usefully be divided into three categories: the early work, in which he translates the anatomical texts of the earlier authorities into pictorial form; later studies based on his own meticulous dissections and observations; and his attempts to extend the form-based observations of anatomy to explore the implications for function – incorporating discussion, analysis and, on occasion, experimentation.

The format of his notebooks is varied. Sometimes he devoted whole pages to a single anatomical form, but more often he undertook a series of studies in miniature, with a larger portion of the page dominated

FIGURE 3.1. Leonardo da Vinci, Anatomical figure showing the vessels and major organs according to Galen, *c.* 1485–90, Italy.

by copious, flawless annotations in his characteristic left-handed, mirror-image script. Interestingly, he seems to revisit certain pages, sometimes years later, to annotate and revise earlier observations. The systematic modern translations of the notebooks into English[14] reveal an astonishing precision and a scope of both description and enquiry that stands far ahead of his time. Around five thousand pages of his notebooks survive; they reveal that he clearly found the heart and its valves most fascinating and worthy of his particular attention. Moreover, these pages capture vividly the evolution of his thinking and overall approach over twenty-five years.

In relation to the heart, Leonardo inherited the conventional wisdom of the day, passed down from Galen and captured in the Arabic texts of Avicenna, to which Leonardo refers in the notebooks. He is known to have owned contemporary textbooks of anatomy, such as that by Mondino de Luzzi (1270–1326). As we learned earlier, in respect of the cardiovascular system, Galen believed that blood was made in the liver, from where it filled the right side of the heart and traversed to the left side through invisible pores: 'the lower ventricles are separated by a porous wall through which the blood of the right ventricle penetrates in to the left ventricle'.[15]

Compare Leonardo's early drawing (Figure 3.1) with any of the twelfth-century English depictions we encountered in Chapter 2 (Figures 2.2–2.4). At first glance, they are very different. Leonardo's version seems so much more realistic: the face has human features, the cross-eyed squint is gone, the torso and limbs have discernible muscular form and he has given the whole an air of artistry and sophistication that is absent from the earlier versions. But on closer inspection, the early renditions and those of Leonardo are all drawn from the Galenic syntheses and have much in common. The liver is prominent as an

FIGURE 3.2 Leonardo da Vinci, The hemisection of a man and woman in the act of coition, *c.* 1490–2, Italy.

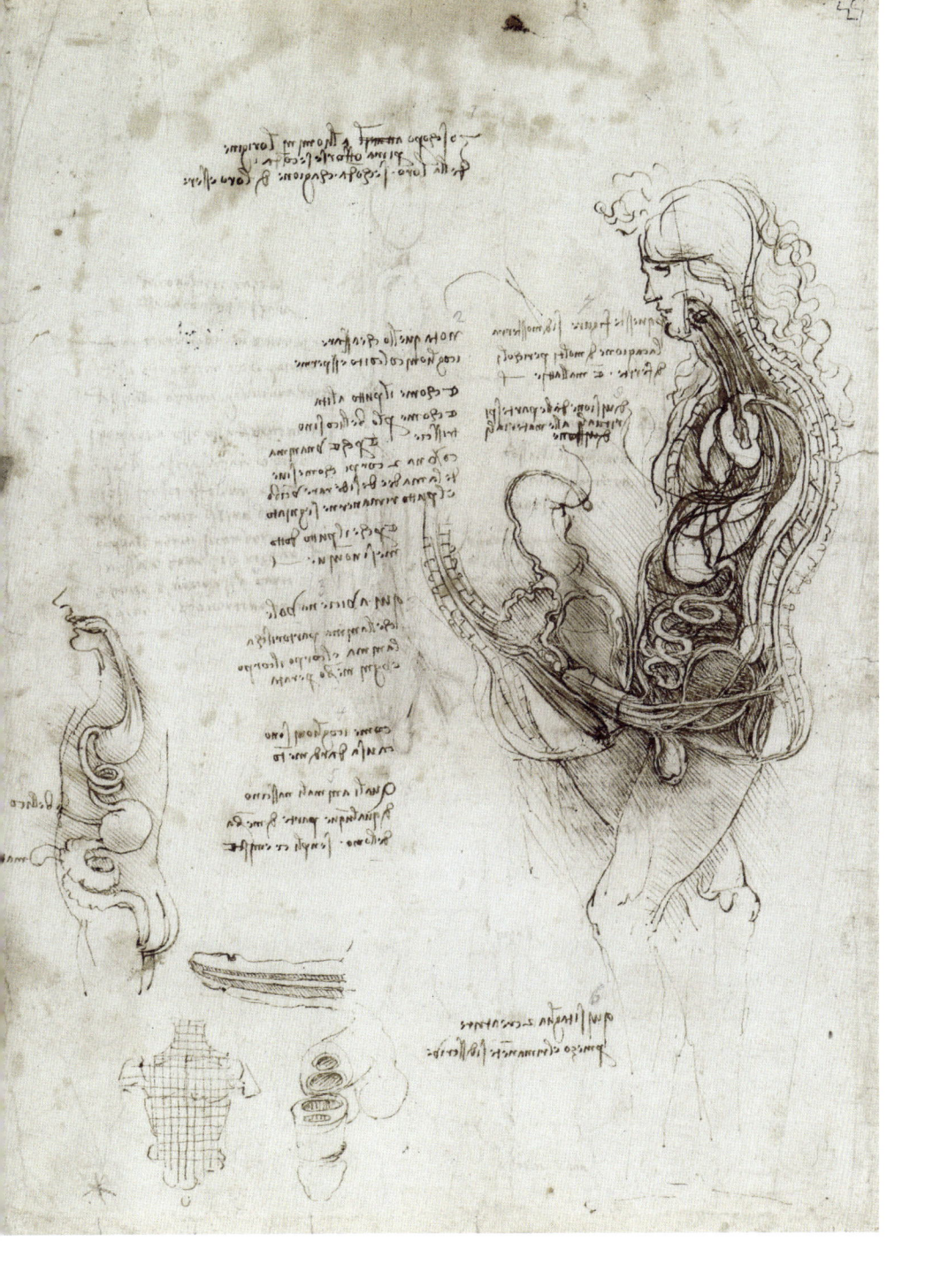

element in the vascular system, the plumbing of the heart is wrong, there are two separate vascular systems and the patterning of the vascular structures has more in common with the early drawings than anatomical reality. All of this suggests that Leonardo not only drew the theory from Galen but may have based his diagrams on the illustrated (probably Arabic) texts. To me the most intriguing and convincing evidence for this interpretation is the two tiny discs that he places within the heart, which are highly reminiscent of the *granum nigrum* and which are without any true anatomical correlate. What did Leonardo imagine they were?

Leonardo scholar Professor Martin Kemp writes that '[Leonardo] clung to the Galenic idea that blood passed through invisible pores in the septum which separates the ventricles.'[16] So, in his earlier work, Leonardo did not shy away from blind depictions of received wisdom. As a result, some of the early 'anatomical' drawings are effectively a fabrication, depicting the learned narratives, albeit with precision and grace. Take, for instance, the prevailing belief that semen was derived from the spinal cord. Plato, in his dialogue *Timaeus*, considered the brain and the 'spinal marrow' as a repository in which 'God implanted his Divine seed': 'And the marrow in as much as it is animate has been granted an outlet from the passage of the egress for drink [the penis].'

In his hemisection of coitus, Leonardo identifies the urethra as a vessel in the penis, but adds in a series of entirely confabulated structures and intricate plumbing that directly link the brain to the penis (Figure 3.2). In fact, closer inspection shows that the sources of semen in this depiction were threefold: the seminal vesicles, the spinal cord and the heart. A tube clearly extends from the heart, along the front of the spine and through the bladder. Eventually it fuses with the tube that seems to have its origin in the testes, although the precise termination of the tube is unclear. Leonardo was undoubtedly draughtsman enough to have offered a clear anatomical course should he have wanted to, but he did not. The ambiguity of the 'plumbing' therefore raises the question of whether Leonardo already doubted the conventional wisdom and whether he was struggling to reconcile what he saw with what he knew.

Nonetheless, at least initially, Leonardo did not feel bound by representations of actual anatomical structures, but instead his early work reflected 'a blend of traditional (and often ancient) beliefs, animal dissections, proportional analysis and mere speculation'.[17]

Clearly, though, Leonardo became dissatisfied with prevailing notions of human anatomy, and he adopted – and defended – a new approach in which he depicted faithfully what he saw, even where this contradicted conventional wisdom:

> Many will think that they can with reason blame me, alleging that my proofs are contrary to the authority of great men held in great reverence by their unexperienced judgments, not considering that my works are the issue of simple and plain experience which is the true mistress.[18]

His revised version of the male genito-urinary anatomy (sometime after 1508) is discussed by Denis Noble and colleagues, who write:

> Leonardo produced the drawing that can be interpreted as his claim to have solved the problem. It must have been based on careful dissection because all the main features of the human male anatomy are correct.[19]

In describing his dissection technique, Leonardo wrote (around 1508–10):

> I have dissected more than ten human bodies, destroying all the organs, and taking away in its minutest particles all the flesh which was to be found around the vessel without causing them to bleed except of the imperceptible bleeding of the capillary vessels.[20]

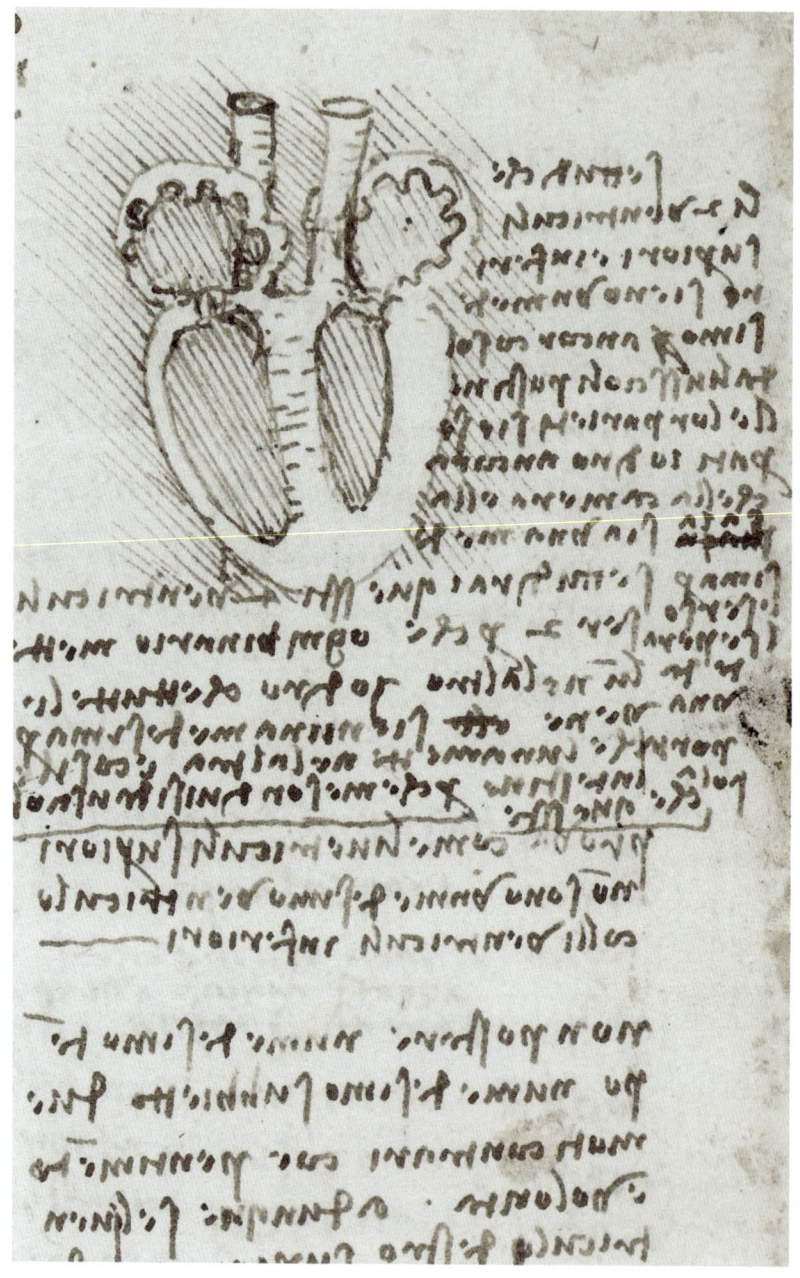

This pattern of investigation is evident in his later studies of the cardiovascular system; early drawings reflected Galenic teaching and, as a result, were simply wrong. Presumably Leonardo recorded Galen's explanations as a starting point for his investigations because this was what was expected of him; synthesizing contemporary beliefs and knowledge into a workable narrative was an approach to which his contemporaries could relate. Modern science is methodologically not so different. It questions, undermines and tests falsifiable hypotheses and, in so doing, approximates ever more closely to an accurate understanding. As we shall see, in these respects, Leonardo can be regarded as a modern, empirical scientist.

But let us return to the 'pores' inside the heart. As I emphasized in Chapter 2, these are not a passing anatomical detail. In Galen's synthesis, blood made in the liver entered the heart through the veins, but that immediately posed a problem. How could blood cross the thick muscular septum that insulates the right and left ventricles? In other words, he needed an explanation of how both arteries and veins could be filled with blood. To solve the problem, he proposed that blood crosses through pores in the septum, thereby linking the two vascular networks. This led to a fundamentally flawed explanation of cardiovascular function that prevailed for well over a thousand years. The problem is that when a narrative takes hold, sometimes decorated with reinforcing myths (in this case concerning humours and spirits), subsequent observations and ideas are interpreted in the context of the parent narrative, or even suppressed because they cannot be made to fit.

FIGURE 3.3 (opposite) Leonardo da Vinci, Schematic representation of the heart, with tentative septal pores, *c.* 1511–12, Milan.

FIGURE 3.4 (overleaf) Annotated pages from the notebooks of Leonardo da Vinci, *c.* 1511–13, Milan.

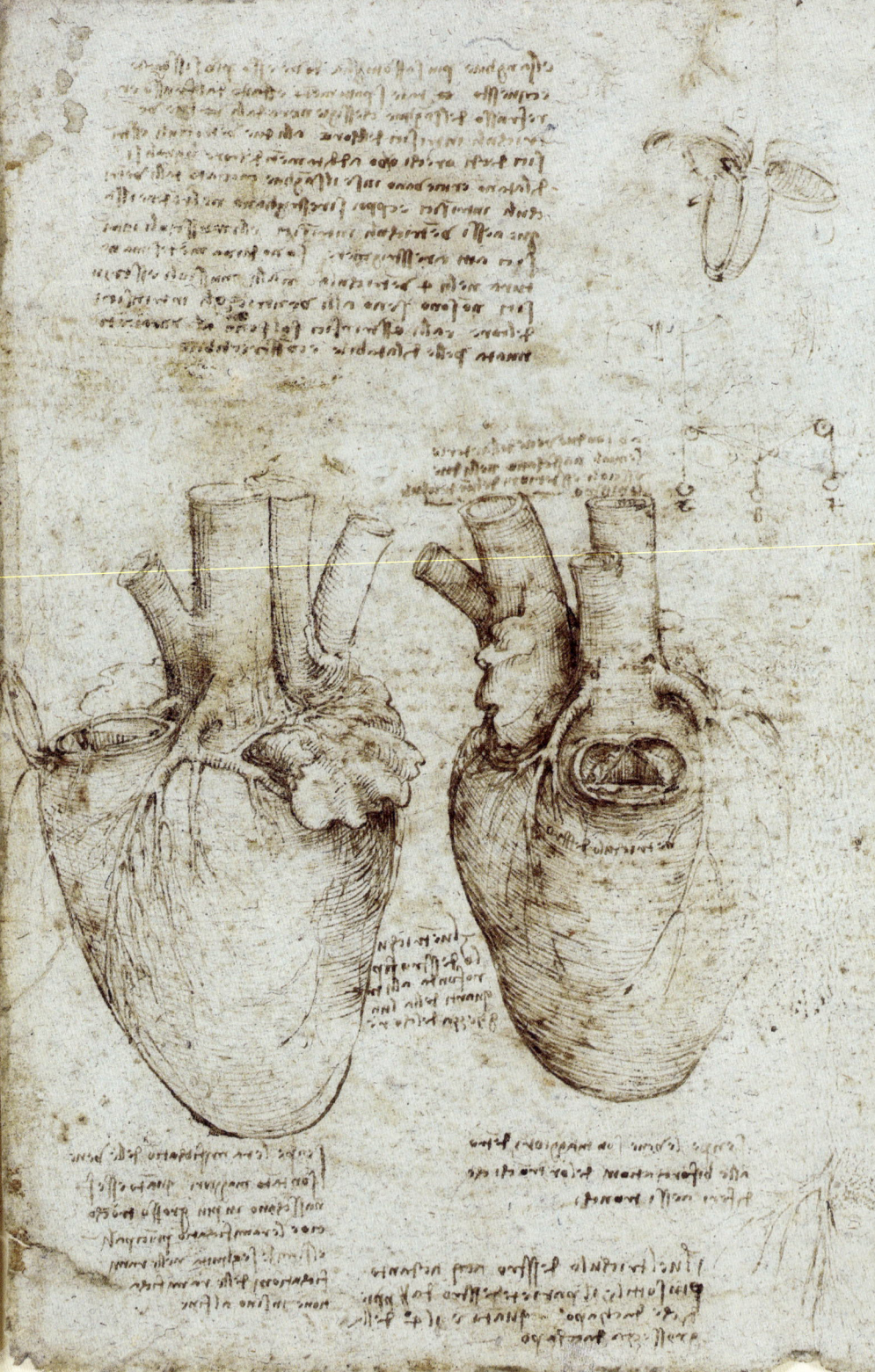

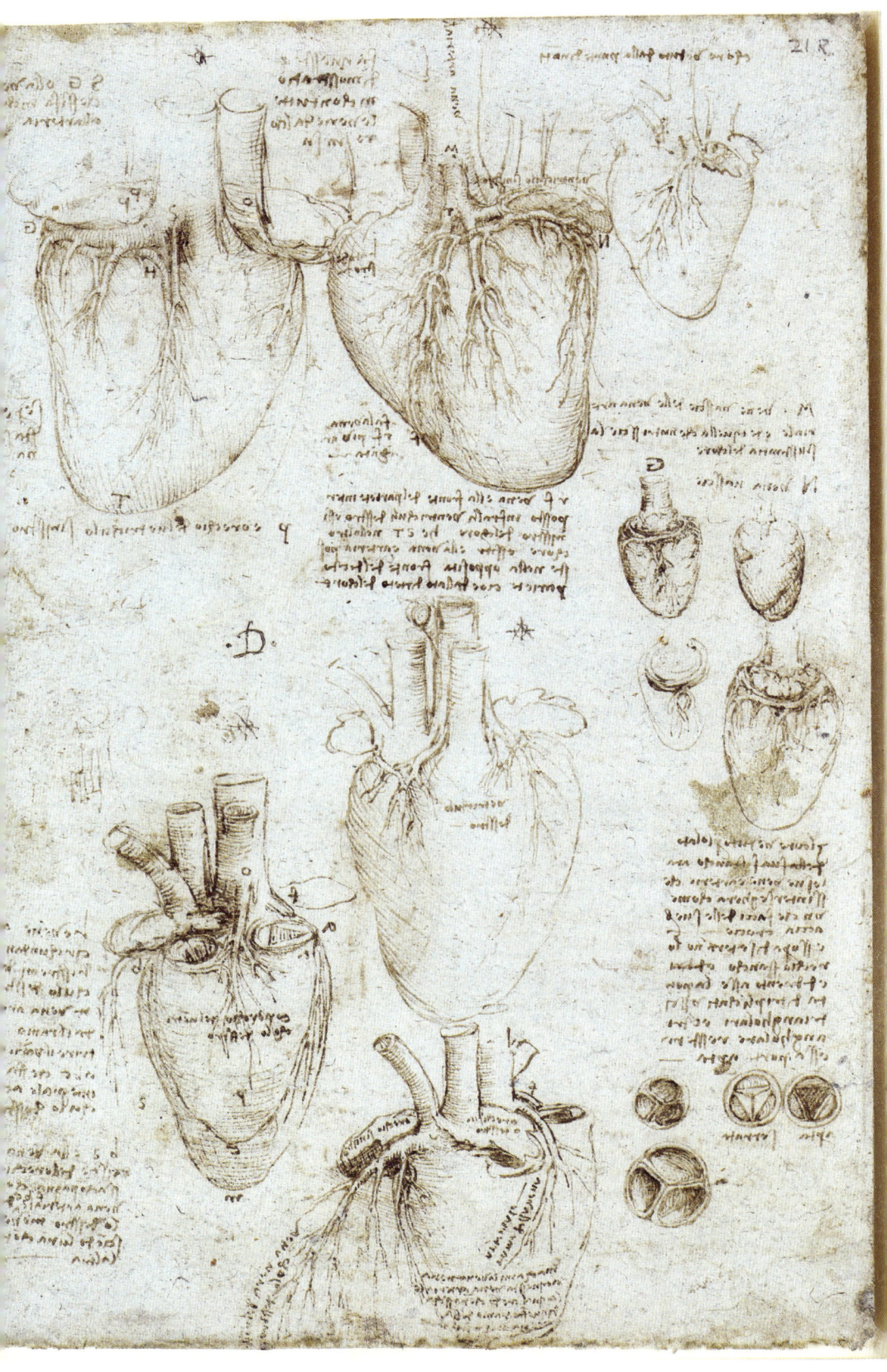

Given that, by 1513, Leonardo's approach was extremely meticulous, paying intense attention to detail, it is hard to believe that he did not take the trouble to examine (or even to test) the presence of these pores before attempting to depict them. He must have known that there were no actual pores in the interventricular septum, or at least that he could not find them. In fact, examined closely, his particular depiction of the pores suggests exactly that. Leonardo's sketch is very 'physiological'; in contradistinction to his truly anatomical drawings, he purposely deconvoluted the architecture of the heart to emphasize its left and right sides, their connections and, by extension, their function. In their true structural form, the aorta and pulmonary artery are intertwined; the aorta loops back over the pulmonary artery as it branches in a complex three-dimensional form, but Leonardo represents them as simple pipes. In the same drawing, he depicts the left ventricle of the heart with its relatively thick muscular wall and the adjacent right ventricle. They are separated by a muscular partition wall: the interventricular septum. It is clear that in his schematic diagram Leonardo *could* have drawn channels traversing the septum and actually linking each ventricle, but he did not (Figure 3.3). Despite Galenic convention, in the structurally normal heart, there is no possibility of direct transmission of blood from one side to the other. In Leonardo's sketch, there is a nod to Galen, but on careful examination none of the tentatively drawn strokes actually crosses the septum from one side to the other; it is a somewhat non-committal gesture. As with the plumbing in relation to the production of semen, there is a strong sense that Leonardo could not reconcile his actual observations of the heart with Galen's theories. He chose, therefore, to depict ambiguity. One can almost sense the intellectual strain that Leonardo must have felt constructing such a fudge.

Why did Leonardo corrupt his otherwise consistently exhaustive and perceptive observation, and even the integrity of his depiction, with subjugation to Galenic orthodoxy? The answer is presumably that, at least at the start of his work on anatomy, he was constrained by the prevailing 'physiological narrative'. Indeed, in other sketches, the septum is depicted as thick, muscular and intact. His later drawings show a clear shift to a more realistic and observed style of anatomical depiction.

Judging by his interpretation of the timing and sequence of the contractions of the chambers in the beating heart, Leonardo must have observed the heart working *in situ* – in the opened thorax of a living creature, probably a pig.

> The upper ventricles [now termed the atria] continually make a flux and reflux of blood which is continually pulled or pushed through the lower ventricles from the upper. And since the upper ventricles are more suited for driving out of themselves the blood which dilates them than pulling it into themselves, Nature has so made it that by the closure of the lower ventricles (which close on their own) the blood which escapes from them is that which dilated the upper ventricles. These through being composed of muscles and fleshy membranes are suitable for dilatation and for receiving as much blood as is pushed into them: they are also suited by their powerful muscles for contracting with impetus and for driving out of themselves the blood into the lower ventricles, of which when one opens the other closes. And the upper ventricles [atria] do the same thing in such a way that when the right lower ventricle opens the left upper ventricle contracts and when the left lower ventricle opens, the right upper one closes. And so, by flux and reflux made with great rapidity, the blood is heated and subtilized and made so hot that but for the help of the bellows called the lungs, which draw in fresh air dilating and pressing it into contact with the coats of the ramifications of the vessels refreshing them, the blood would become so hot that it would suffocate the heart and starve it of life.[21]

In other words, Leonardo, who was ignorant of the *circulation* of the blood, believed that it moved back and forth within the heart, encountering friction as it moved over the irregular inner surface, thereby providing a mechanism for the generation of heat. Indeed, he went further, speculating that this friction would lead to overheating, were it not for the cooling effects of air, with which blood came into contact through the branching blood vessels in the lungs. His parenthetic

FIGURE 3.5 Leonardo da Vinci, Studies of the vortices in flowing water, *c.* 1509–11, Milan.

comment that the ventricles 'close on their own' alludes to the automaticity of cardiac contraction and anticipates one of René Descartes's responses to William Harvey's elucidation of the circulation of the blood.

By 1508, Leonardo had moved from Florence to Milan, where he was to remain until 1513. This was a time when Leonardo engaged in wide-ranging scientific and technological studies, embracing mathematics, optics, geology, botany and hydrology. His professional responsibilities included surveying locks and water channels (an interest that Vasari traces back to his early life); he was fascinated by the movement of fluids. He combined these hydrological investigations with intensive anatomical studies. In particular, he was struck by the similarities between water flows past obstacles, such as lock gates, and the passage of blood across the heart valves. The jump from observations of the eddies and vortices of water flowing through rivers and irrigation channels to the flow conditions of blood in the aorta is, perhaps, not so remarkable for Leonardo. His worldview was one of integration of the design of nature and human design, and it would have nudged him towards a unifying understanding, in which the human 'microcosm' did not sit apart from the processes of the natural world but conformed to its order and operated according to its rules.

From the 1490s, Leonardo and others moved away from metalpoint drawing and began to develop the use of chalks (red and black) for a range of graphic effects.* In his series of water studies, the chalk effectively conveyed the mass and force of water and, most significantly, its flow (Figure 3.5). Moreover the contrast created by the use of red chalk – for example, in his drawing of the human foetus – had the additional effect of conveying some sense of tissue, and even vitality.[22]

---

* Pencils today use graphite, which became popular as a material for drawing towards the end of the sixteenth century. In the Renaissance period, drawings were made using a variety of fine metal tools known as 'metal points'. They enabled fine work that lent itself to the more objective styles of representation found in most of Leonardo's anatomical drawings.

Leafing through Leonardo's copious anatomical studies, I can't help but see the studies of the valves of the heart as a high point in these works. The meticulous drawing and annotations exemplify his willingness to shed received thinking, refocusing on intensive observation and demonstrating a transition from over-reliance on static structure and connection towards dynamic function. He went over the valves again and again, studying them from all angles and imagining the flow of blood through their pliable leaflets. His many anatomically faithful cross-sections emphasized the functionally important components and reflected a completely new approach. Combined with the integration of – and extrapolation from – form to function, this represented a major conceptual leap. Leonardo had moved from the structural discipline that is anatomy to its functional cousin, physiology, and extended his ideas by developing experimental model systems, as we shall see below.

Around 1512–13, Leonardo studied both the mitral and aortic valves in extreme detail. But it is in his systematic examination of the aortic valve that we can see his virtuosity most strikingly. As his starting point, Leonardo noted the structural similarities with the gates of canal locks, which created disturbances in the flow of river water. He even termed the valve the 'little gates'. These observations led him to describe, by analogy, 'layered' vortices that he postulated occurred in the small bulges in the proximal aorta (now known as the 'sinuses of Valsalva'; had Leonardo published his findings, they would surely have been named after him). He describes again and again, in sketches, the patterns of blood flow across the valve and the precise anatomical location of the vortices of blood in the sinuses just beyond the valve (Figure 3.6).

To test his theory experimentally, he constructed a glass model built from a cast of the aortic root of a cow. Leonardo wrote: 'pour the wax into the gate of a bull's heart that you may see the true form of this gate'. From this cast, he built an anatomically faithful glass model, into which he introduced grass seeds, suspended in water, that allowed him to map the flow vortices. Based on his hydrological studies, he had postulated that blood in the centre of the stream would flow faster than that impeded by friction at the vessel edges. It is far from obvious that this would or could happen, and the extrapolation from structure to function is startlingly ahead of his contemporaries. Even more remarkably, Leonardo not only identified the existence of the vortices but also calculated the physiological advantage they would confer. To this day, medical schools wrongly teach that the incipient flow reversal of blood at the end of systole (heart contraction) is due to pressure changes, when blood has been ejected and the heart is beginning to expand, leading to the closure of the valves during diastole, the period when the heart is filling with blood. But, for Leonardo, the purpose of the eddies of blood in the sinuses at the end of systole was to effect active closure of the aortic valve, as an integral consequence of the end of the previous heartbeat. Of course, unlike the water in the gates of the locks, it is not possible to see these vortices – even were the aorta transparent, it is inconceivable that these vortices could be discerned. So, Leonardo was capturing an imagined phenomenon. His deductions are all the more remarkable because he had no notion that the entire cardiac output exited the heart to be replaced by blood entering the heart on a constant circulatory loop. In the passage above, it is clear that he thought most of the blood moved to and fro within the heart. In relation to flow across the aortic valve, he refers to the 'small' amount of blood actually ejected from the heart.

The question of whether Leonardo was right in his assertions about the aortic valve has intrigued modern researchers. A paper published in the magazine *Nature* in 1968[23] recreated Leonardo's model and confirmed his interpretation that the eddies of blood were important in assisting the timely closure of the aortic valve. Were the eddies not to perform this role, the functions of the valve would be severely compromised, allowing substantial backwards leakage of blood from aorta to heart after every heartbeat, due to delayed closure of the valve.

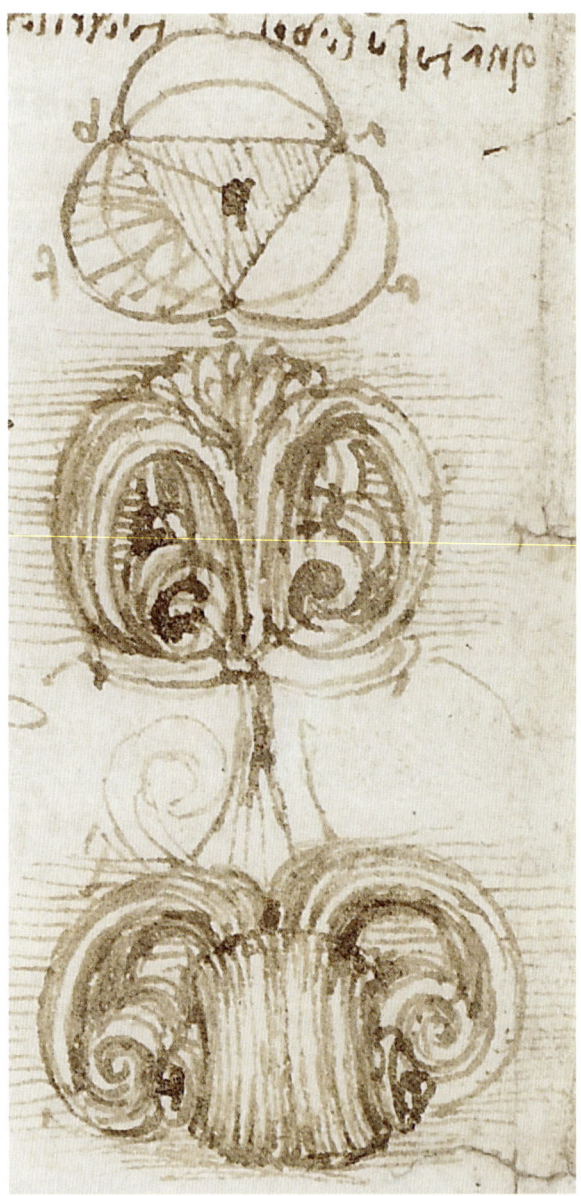

FIGURE 3.6 Leonardo da Vinci, Geometries of the three aortic sinuses and the formation of vortices in flowing blood, *c*. 1512–13, Milan.

Nonetheless, an actual demonstration of the vortices in a living animal remained elusive until very recently, after scientists at Stanford University reported an astonishing new method that they termed 'four-dimensional (4D) magnetic resonance velocity mapping'. In short, this non-invasive technique uses radio-frequency waves to tag blood with high spatial resolution and to follow both its speed and direction over time (the fourth dimension). Until then, it had been possible to use computational techniques to model the flow of blood, but with the advent of 4D-flow mapping, it became possible to visualize the actual patterns of blood flow. My first reaction was to wonder at the ingenuity of the technology and the beauty of the moving images; my second was to realize that we might use 4D-flow mapping to 'see' if the vortices were truly there and to test Leonardo's hypothesis on a living human. In 2013, exactly 500 years after Leonardo's original observation, my laboratory in Oxford had the opportunity to do just that.[24]

The correspondence between Leonardo's 500-year-old theory and our findings was remarkable. The image (Figure 3.9) shows six sequential, velocity-encoded images of blood flow in the aortic root, appearing as cine frames of a single heart contraction (systole). In the early phase, there is a laminar (smooth, undisturbed) flow across the width of the vessel through the open aortic valve. Gradually the flow pattern becomes differentiated, with the generation of vortices at the edges of the aorta. These occur in the precise location that Leonardo predicted, in the same size and proportion as his drawings and, crucially, they occur only at the end of systole, all of which is entirely compatible with his theory of the function of the vortices in effecting valve closure. The composite figure shows the close similarity between postulated and measured blood vortices in the aortic root.

FIGURES 3.7 AND 3.8 (overleaf and following spread) Pages from the notebooks of Leonardo da Vinci, *c.* 1511–13, Milan.

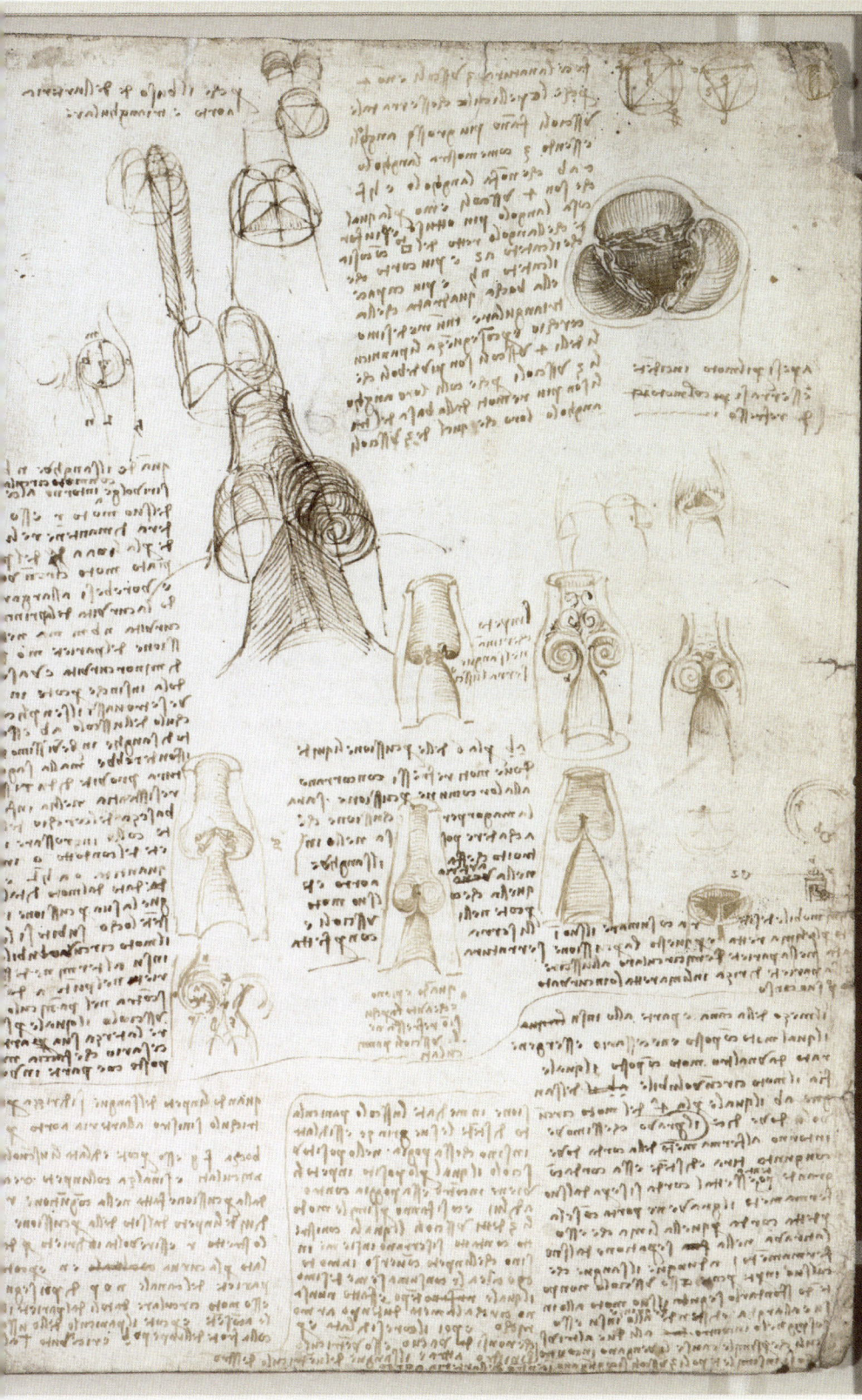

Interestingly, the secondary inwardly curving eddies that Leonardo showed above the level of the aortic sinuses, which would not be necessary for valve closure and which seem not to have been generated by his experimental observations, are not present in the 4D-flow image. Still ensnared by false narratives and ignorant of the circulation of the blood, Leonardo conjectured that these vortices were involved in the dissipation of the kinetic energy of blood and, thereby, heat.

Over the twenty-five years of his anatomical studies, Leonardo demonstrated, using an entirely new style of representation, a deep appreciation for and systematic analysis of anatomy. Had his work been published in his time, the field of study would have developed differently and our perceptions of the man and his work would almost certainly have been very different. Given the huge advances he achieved through the virtuosity of both his descriptions and his interpretations, I find it hard not to feel some sense of intellectual frustration on Leonardo's behalf that he could not crack the crucial central fact of the circulation of blood, and instead was limited to ever more detailed descriptions of the components of the cardiovascular system. This is all the more frustrating given the evidence that he recognized some of the limitations of the Galenic syntheses, and that he recognized many of the key elements that necessarily imply the circulation of blood. He understood the function of valves, recognized the heart as the centre of a branching vascular system and appreciated the passage of blood from large vessels to tissue-level capillaries. But an elucidation of the circulation of the blood had to wait for a little over a century, when William Harvey, without materially important new information or technology, was to crack arguably the most fundamental question in biomedicine.

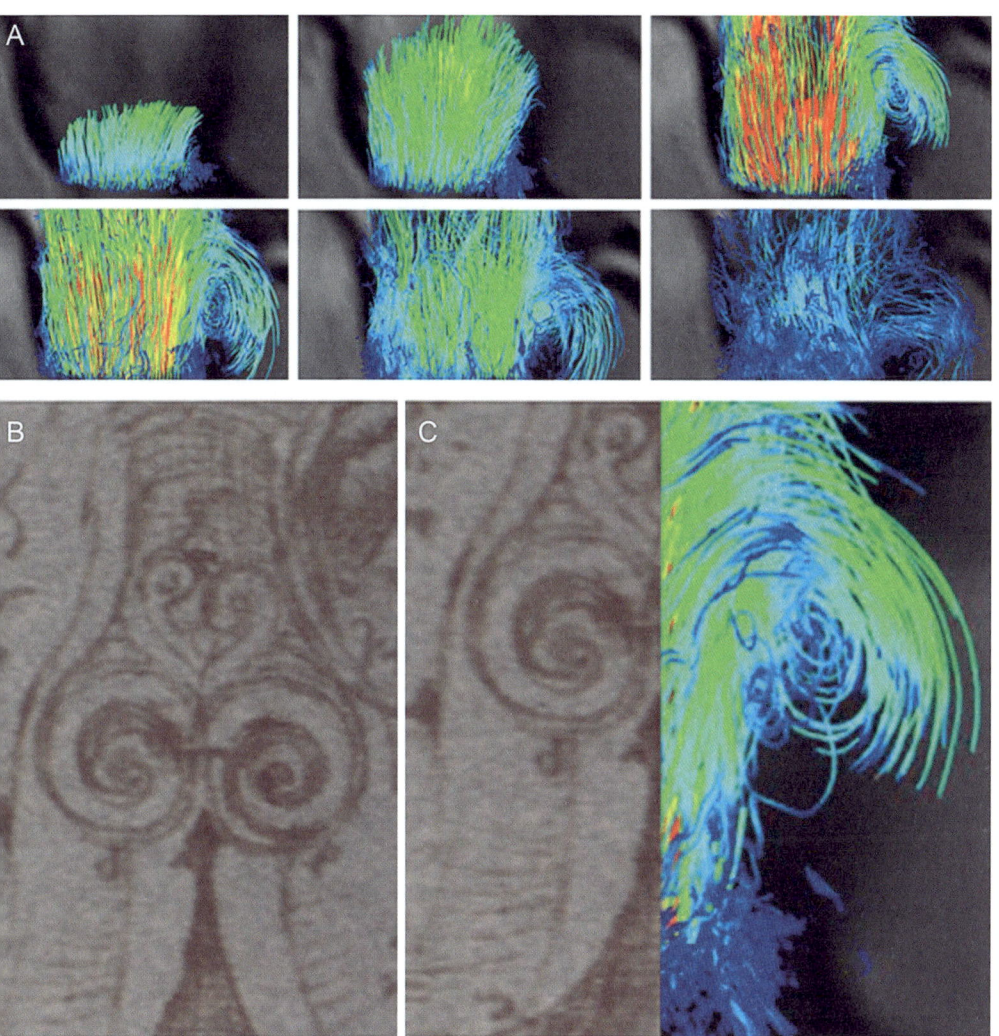

FIGURE 3.9 Twenty-first-century mapping of blood flow in the aortic root using magnetic resonance imaging techniques, 2013, Oxford.

WILLIAM HARVE. M.D.

William Harvey and the Impact of Circulation

To the Most Illustrious and Invincible Monarch
Charles King of Great Britain, France and
Ireland, Defender of the Faith.
Most Gracious King,
The Heart of creatures is the foundation of life,
the Prince of all, the Sun of their microcosm, on
which all vegetation does depend, from whence all
vigour and strength does flow. Likewise, the King
is the foundation of his Kingdoms, and the Sun of
his microcosm, the Heart of his Commonwealth
from whence all power and mercy proceeds.

WILLIAM HARVEY
De Motu Cordis, 1628

So begins the dedication (to Charles I of England) of *Exercitatio anatomica de motu cordis et sanguinis in animalibus*, more usually known as *De Motu Cordis* (*On the Motion of the Heart*), arguably the most important treatise in the history of medicine. William Harvey left little room to doubt his view of the primacy of the heart, or its role as the provider of life. In Harvey's analysis, the heart does not depend on any other organ – it is itself the life source and originator. In this regard, Harvey aligned firmly with Aristotle, of whom he was a declared follower.

In researching this book, a large part of my motivation (as a cardiologist dealing not only with the heart but with people's attitudes to their hearts) has been a curiosity to understand how and why the heart has attained its position as the pre-eminent organ – not merely in the physiological sense, but as the object of persistent reverence and fascination that go far beyond its prosaic role as a mechanical pump that Harvey and his immediate successors were about to reveal. I find it telling that at the very moment of his revelation of the function of heart, Harvey elevated it further through metaphor, rather than attempting to diminish it with his new knowledge or presumed intellectual mastery.

In fact, Harvey's grandiose regal metaphors belied his actual reticence in publishing his highly disruptive text. In general, endeavours in medicine up until the mid-sixteenth century were not driven by a desire to discover and innovate, in our contemporary sense, but to demonstrate knowledge and the application of long-established classical texts. In that context, Harvey had profound doubts that his work would be well received.

Recall that in Galen's synthesis, blood was made in the liver and passed on to the heart through the veins. From there, it ebbed and flowed to the tissues, where it was consumed; a small amount crossed the septum of the heart and became 'vivified' by air from the lungs. Although wrong, Galen's explanation was a workable narrative with a certain internal consistency; it was not by mere chance that his ideas were so pervasive or persisted for over a thousand years. Galenic thinking also underpinned the widespread and lucrative practice of bloodletting, which was prescribed for a variety of maladies, but which would not sit well with Harvey's demonstration of the circulation of a relatively small and finite blood pool in a closed system. By extension, the ancient theories of disease built around the imbalances of the four humours were also undermined. So, although Harvey was not alone in questioning the Galenic conventions, he was still clearly nervous about the prospect of expounding a complete and radical revision. He refers to the nine years in which he demonstrated the relevant anatomy and physiology underlying his theories to his closest colleagues, the Fellows of the Royal College of Physicians:

> I did open many times before… my opinion concerning the motions and use of the heart and circulation of the blood in my lectures; but being confirm'd by ocular demonstration for nine years and more in your sight, evidenced by reasons and arguments, freed from the objections of the most learned and skilful anatomists, desired by some and most earnestly required by others, we have at last set it out open to view in this little Book…
> I can call most of you, being worthy of credit as witnesses of those observations from which I gather truth, or confute error, who saw many of my dissections, and in the ocular demonstrations of these things which I here assert to the senses were us'd to stand by and assist me.[1]

So, Harvey was both declaring the radical nature of his elucidation and taking cover by reference to the sustained attention, and implicit agreement, of his colleagues, who witnessed and participated in the dissections and demonstrations that confirmed his theories:

contrary to the received way through so many ages of years insisted upon and evidenced by innumerable, and those most famous and learned men, I was greatly afraid to suffer this little book.

However:

True philosophers, who are perfectly in love with truth and wisdom, never find themselves so wise or are so full of wisdom or so abundantly satisfied in their own knowledge, but that they give place to truth whensoever all from whosoever incomes.

Harvey goes on in this vein for several pages. In spite of (or perhaps because of) his elevated positions in the Royal College of Physicians, as the king's physician and as physician to the oldest and largest of the London hospitals (St Bartholomew's), his anxiety is palpable.

It is remarkable that in his elucidation of the circulation of blood Harvey brought no new information and certainly no new technology to bear – his remarkable accomplishment was born from rigour, rational analysis and a willingness to take considerable professional risks in challenging the prevailing view. His approach was undoubtedly fuelled by the intellectual environment in which he operated. Shortly after Harvey made his incisive and disruptive observations, his sometime patient Francis Bacon published (in 1620) his attempt to provide some formal shape to the experimentally driven science that was then emerging. Bacon's *Novum Organum* (*The New Organon*) takes its title from Aristotle's works on logic, the *Organon* ('instrument for rational thinking'). Bacon argued that Aristotelian logic was unsuitable for the pursuit of knowledge in the 'modern' age and proposed a reason-based system, suitable for the pursuit of *knowledge*:

Where Aristotle's inferential system based on syllogisms could reliably derive conclusions which were logically consistent with

an argument's premises, Bacon's system was designed to inves-
tigate the fundamental premises themselves. Aristotle's logic
proposed certainty, based on incontrovertible premises accepted
unquestioningly as true; Bacon proposed an inductive inference,
based upon a return to the raw evidence of the natural world.[2]

The observational method proposed by Bacon in *Novum Organum*
was based on what he described as the 'fresh examination of particulars'.
Systematic analysis and fresh interpretation were precisely the elements
of investigation that William Harvey applied to elucidate the function
of veins and arteries and their relationship to the heart. He described his
findings in seventeen short, elegantly structured chapters, in which he
dismantled 1,400 years of received wisdom and laid the foundation for
modern medicine, surgery and therapeutics. I do not intend to imply
that Harvey's approach was derived from Bacon; indeed, as Thomas
Huxley was at pains to stress in his essay of 1878 published in the
Fortnightly Review:

> In the latter half of the 16th century and the beginning of the
> 17th centuries the future of physical science was safe enough in
> the hands of Gilbert, Galileo, Harvey, Descartes and the noble
> army of investigators who flocked to their standard, and followed
> up the advance of their leaders. I do not believe that their won-
> derfully rapid progress would have been one whit retarded if the
> 'Novum Organon' had never seen the light; while if Harvey's
> little 'Exercise' had been lost physiology would have stood still
> until another Harvey was born into the world.[3]

Radical as they were, Harvey's deductions did not occur in isolation
– his work evolved from the mainstream of medical endeavour that
had its roots in the transmission of Greco-Arabic learning into Europe,
largely through the Latin translators in Spain, from the middle of the

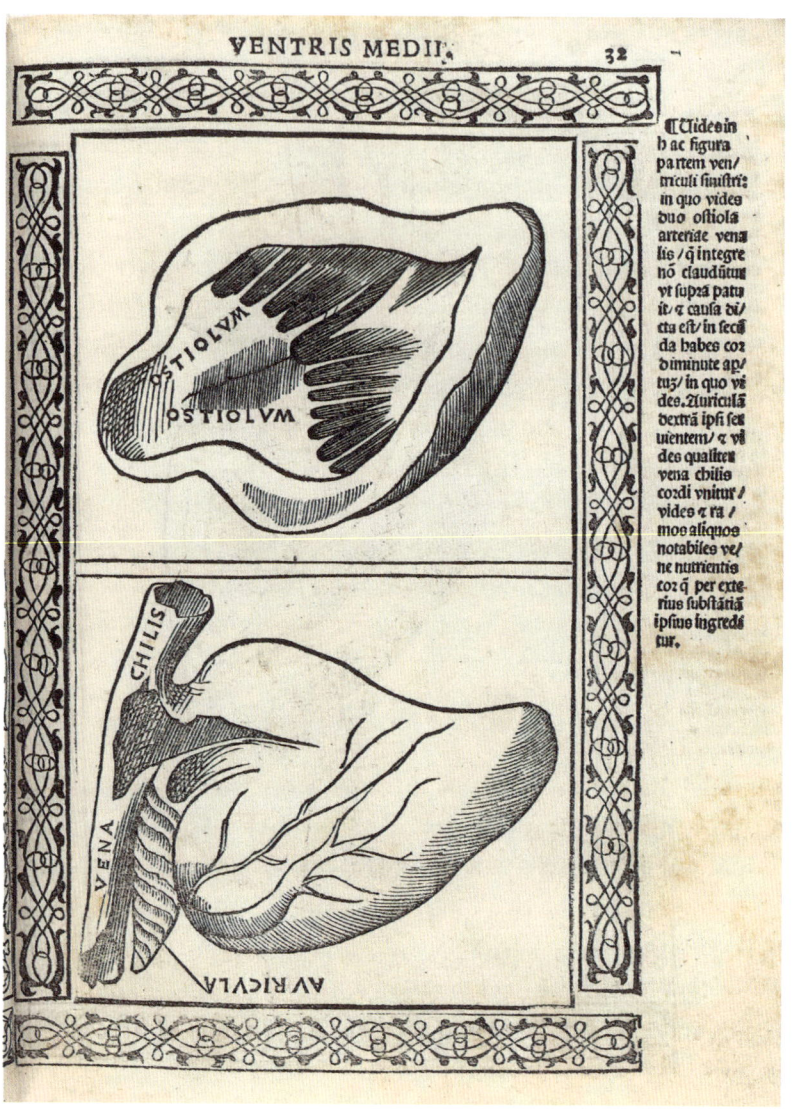

Figure 4.1 Jacopo Berengario da Carpi, *Isagogae breves in anatomia*, 1523, Bologna.

twelfth to the middle of the thirteenth centuries, and through Salerno in Southern Italy. With the emergence of the universities, particularly in Bologna (also in Paris and Montpellier), anatomical study became organized and communicated. Mondino de Luzzi, a professor in Bologna (1306–26), instituted human dissection (remember Galen and Aristotle had dissected only animals) in a more formal sense than had been undertaken previously,[4] and in 1316, Mondino wrote his text on anatomy. When it was first printed in Pavia in 1478, the only illustrations it contained depicted the process of dissection. Study of Mondino's text – a landmark in the medieval rediscovery of human anatomy – marked the beginning of structured medical tuition in Europe, focused in the universities and catering for international students. In 1508, a collection of Galen's works was translated directly from Greek into Latin by the Italian physicist and humanist Niccolò Leoniceno (1428–1524), and was published in Paris and received avidly. Leoniceno incorporated in his translation a critical evaluation not only of the fundamental tenets of Galenic medicine but also of earlier translations of his work. Leoniceno's approach was not so much one of modern scientific discovery as one of 'trust but verify'.

But the sixteenth century was about to become the century of anatomy. Anatomy was attracting wider interest, and public dissections of executed criminals were well attended.[5] Northern Italy became the epicentre of this activity – in the universities of Padua, Bologna and Pisa. In relation to depictions of human anatomy, a key shift was the move from written text to printed manuscript and the possibility of reproducing figures at scale using the skills of draftsmen, engravers and printers. In the early days of this exploration, and for a good portion of the Renaissance period, artists and printers played a prominent role – in essence reporting how the structures appeared, but by implication without deeper understanding of the context or function of what they were portraying. By the middle of the sixteenth century, anatomists had taken over as the primary depicters. In his history of anatomical illustration,

Choulant credits Berengario da Carpi as the first depicter of anatomical images drawn from observation.* In case any more evidence were needed of Leonardo's distinction as an anatomist, contrast the crudeness of the Berengario woodcuts (Figure 4.1) with the extreme sophistication of Leonardo's drawings from almost precisely the same time.

Johann Eichmann (called Dryander, 1500–60), professor of medicine in Marburg, was another early anatomist to draw from his own dissections. His 1537 publication contains twenty plates, mostly of the head and neck, but the last four plates are of the heart and lungs.[6] Here (Figure 4.2) we see the posterior aspect of the heart in a highly stylized form, to the point of appearing almost like a plucked fruit presented in front of a landscape. The primary interest of the engraver is unlikely to have been anatomical depictions.[7] It was common for anatomical works to be undertaken as part of the training of engravers, and one imagines that many might have been more interested in placing a virgin and child, or Venus rising from a clam shell, in the foreground, rather than this dry anatomical subject.

On the left we see the heart from behind. The tubular structure with a trumpet-like mouthpiece curving from the left and piercing the diaphragm is the oesophagus, which courses through the chest behind the heart. Sitting immediately behind the oesophagus is the trachea, encased in rings of cartilage. It seems more than redundant coincidence that Dryander included the trachea in his depiction of the heart. Although this may simply reflect their anatomical proximity, there seems more than a hint in the Dryander that the trachea is adherent to, perhaps even plumbed into, the heart, reflecting the prevailing Galenic thought that the air, or vital spirit, was mixed with

* Ludwig Choulant published his *History and Bibliography of Anatomical Illustration* in Dresden in 1851. With an emphasis on the history of pictorial description in anatomy, he summarized the contributions of around a hundred European anatomists and included an appendix on Chinese anatomy. Besides his work as an historian of anatomy, he was the author of a textbook of internal medicine.

blood in the left ventricle of the heart. Even the briefest comparisons with Leonardo's almost exactly contemporaneous drawings (if anything, Leonardo was earlier) show the gulf in the rigour of their enquiry. Accepting that the printed images are woodcuts and the opportunity for detailed depictions was accordingly less than in Leonardo's ink drawings, there is a clear intellectual distinction between Leonardo and his contemporaries that speaks to the precision of both the former's enquiry and his understanding. His work is distinct from, and stands above, mere depiction.

Bartolomeo Eustachi (1500–74) is most enduringly known for his description of the Eustachian tube (which joins the nasopharynx to the middle ear). His copperplate engravings of the heart are believed to have been made by Giulio de Musi of Rome, who was better known for his architectural works, and possibly accounting for the difference in style compared with the Dryander depictions. Here, the heart and vessels (Figure 4.3), although still highly stylized, are isolated specimens, giving a much crisper, more studied and 'scientific' appearance. This impression is reinforced by the numbers in the margin, which provide page coordinates, similar to those on maps, so that the reader can more readily identify structures described in the text.

The origins of Harvey's discovery

The origins of Harvey's approach and the substance of his enquiry grew from the work he undertook in Padua, which at that time was the pre-eminent school of medicine. In 1600, Harvey began his studies there under the professor Andreas Vesalius (1514–64), though his main teacher was Girolamo Fabrizi (1533–1619), known as Fabricius, from

FIGURE 4.2 (overleaf left and right) Johann Dryander, Representations of the heart (front and rear) from *Anatomia Mundini*, 1537, Marburg.

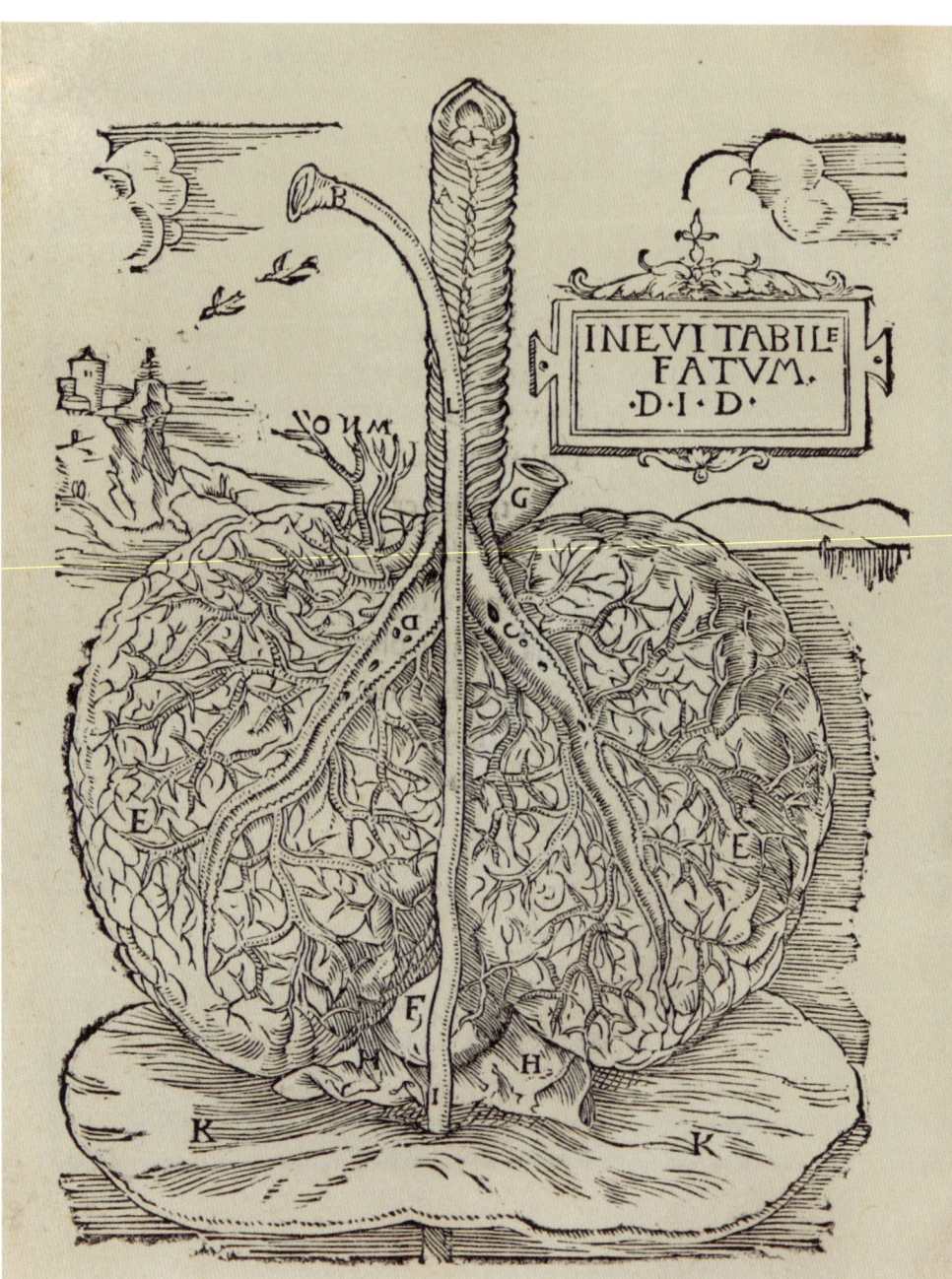

Cordis, eiufdem̃q̃ duorum uen   trium cum uẽnis & arterĩs inde
prorumpentibus particulis, bi-   nosq̃ infignes tramites hunc cibo
atq̃ potui,illũ aeri ppetuo defti-   natũ &c,figura præfens exhibet.

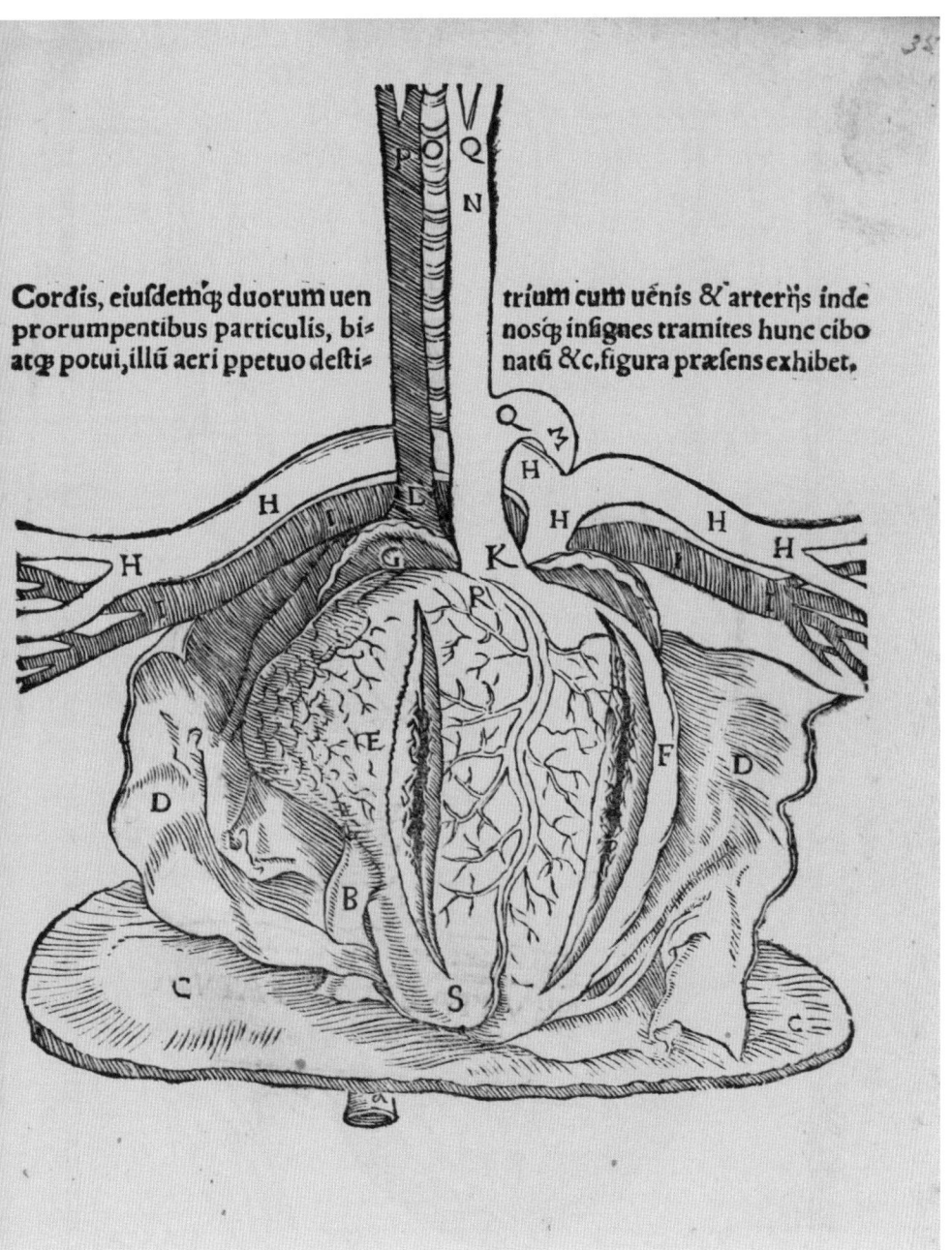

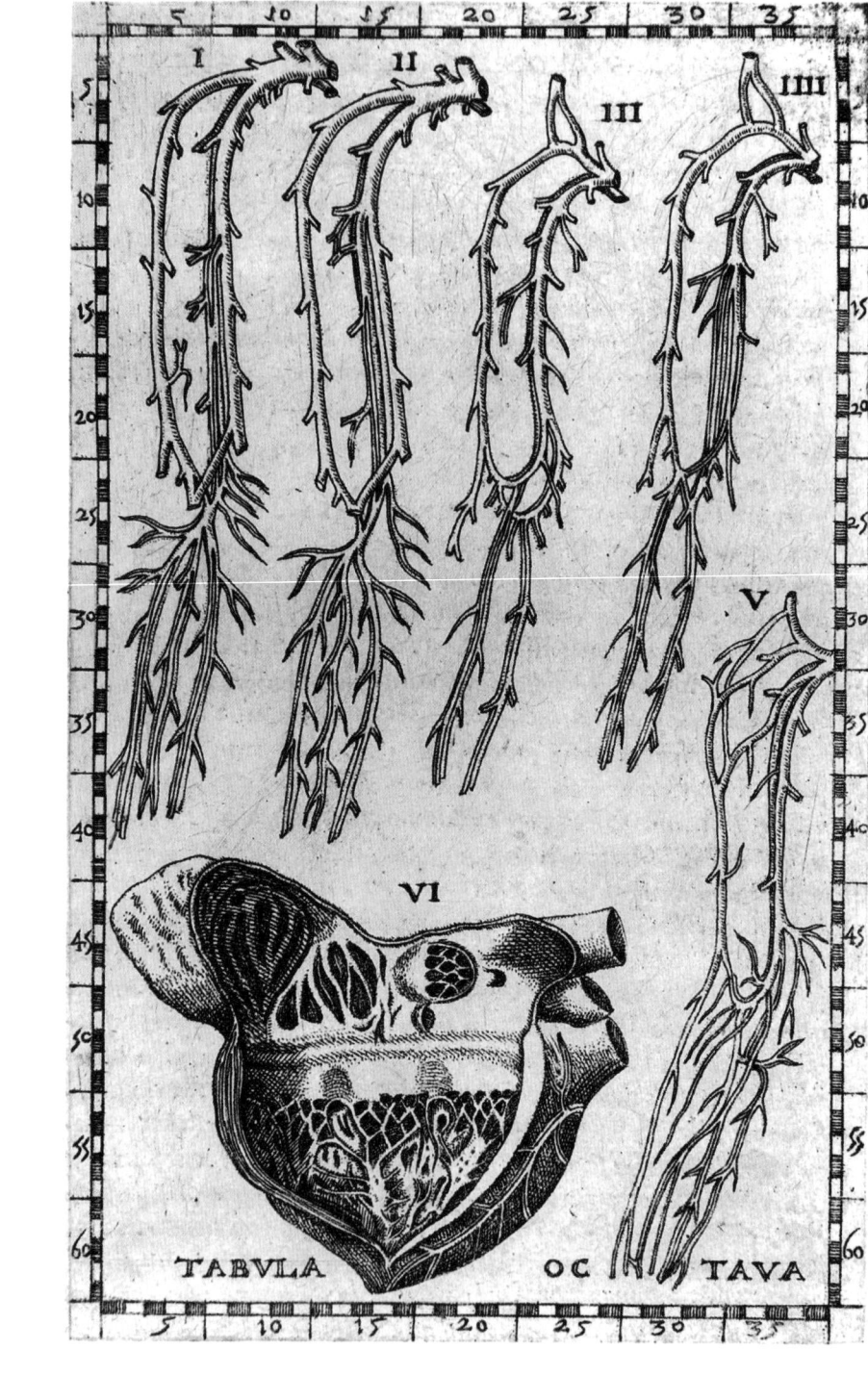

TABVLA   OC   TAVA

whom he developed an Aristotelian approach to the study of nature, especially comparative anatomy and embryology.[8]

Vesalius was a pivotal figure in the study of human anatomy. He distinguished himself early and was elected professor of anatomy and surgery in Padua at the age of just twenty-three.[9] A disruptive thinker, Vesalius became a key contributor to the scrutiny of the old ideas, opening the door to rational understanding through 'hands-on' morbid dissection and meticulous observation. His ambition was an exhaustive description of human anatomy: the morphology of the organs and tissues and their interrelationships and topology. Our contemporary equivalent might be the promise, revelation and unleashing of potential that has accompanied the mapping of the human genome.[10]

In 1595 in Padua, at his own expense, Fabricius had erected a circular gallery to enable observation of dissections, under his direction, with subjects brought up by an elevator from the preparation room beneath.[11] The gallery is surprisingly small; the dissection table occupies almost all of the ground level, with just enough room for the dissector to walk around it. The viewing gallery comprises an oval funnel of steep wooden tiers. There are no seats; to maximize viewing potential, there is standing room only. It is not hard to imagine tightly packed physicians and students from around Europe peering and craning to glimpse the various organs and vessels revealed by the celebrated early masters of anatomy.

As for all contemporary students of anatomy, the starting point for the Padua school was the work of Galen, who believed that blood bathed the tissues, with no notion of continuous, high-volume flow or circulation of blood.[12] Integral to Galen's explanation of heart function were the imaginary pores that connected the right side of the heart (receiving blood from the liver) and the left side (connected to, and receiving air

FIGURE 4.3 Bartolomeo Eustachi, Anatomical depictions of the heart and branching blood vessels from *Opuscula anatomica*, 1564, Venice.

from, the lungs). Without these pores there was no explanation for the presence of blood in both the arteries and the veins. Since the veins and the arteries were separate branching systems, according to Galen, the pores in the heart were necessary to allow blood to move from the veins to the arteries. While using Galen as the starting point (there was no other), Vesalius had not 'indiscriminately accepted all his opinions… and demonstrated some fault is discernible in his books'.[13]

His major published work was *De humani corporis fabrica* (1543). In the first edition of *De fabrica* he hinted that he doubted that the blood passed through the septum; in the second (1555) he stated that he could not find any channels from one ventricle to the other:[14]

> Not long ago I would not have dared diverge a hair's breadth from Galen's opinion. But the septum is as thick, dense and compact as the rest of the heart. I do not see therefore how even the smallest particle can be transferred from the right to the left ventricle through it.[15]

In this respect and others, Vesalius had clearly grasped that the ancient ideas were fallible, and that detailed, empirically gathered knowledge would be essential to a full understanding of human anatomy. But despite his meticulous descriptions, his analysis of the implications of his findings in relation to the heart fell short:

> So far as can be determined by the senses none of these pits goes through from the right ventricle to the left and we are therefore compelled to marvel at the Creator's clever device by which blood oozes from the right ventricle to the left through invisible channels.[16]

De fabrica reflected Vesalius's systematic examination, via dissection, of the whole human body, which he depicted in words and images.

Vesalius sat at the pivot point of a dramatic transformation in the approach to anatomy that took place in the first fifty years of the sixteenth century, as illustrated in depictions of dissection in progress (Figure 4.4).[17] The woodcut on the left is taken from the first Italian edition of Ketham's compendium of anatomical works, *Fasciculus* (Venice, 1494), at the beginning of the section on the *Anathomia* (*Anatomy*) of Mondino. This famous image shows a technical dissection. The disengaged professor is seated behind a lectern, removed from the actual dissection, reading *ex cathedra* from the established texts of Galen or Mondino. The dissection was directed by a junior colleague, the *ostensor*, seen with a baton on the right, while the actual dissection was undertaken by a technician stooping over the corpse, with sleeves rolled up and brandishing a somewhat indelicate hunting knife in his hand.[18] The picture is hardly one of carefully studied dissection and, in reality, the procedure might not have involved much more than incising the body to lay open the contents. The remaining figures are distant from the action and seem utterly uninterested in the proceedings.

Contrast this with the frontispiece to Vesalius's *De fabrica*, which depicts a completely different scene. Here, an array of tightly packed, chattering observers crushes towards the cadaver to learn directly from the observations made during the dissection. Two figures lean out from the gallery, a head peeps between the thigh bones of the suspended skeleton, while various precariously positioned figures, and even a monkey, perch on the periphery. Vesalius himself stands adjacent to the body, with the extended fingers of his right hand drawing attention to some feature in the gaping abdomen, while he holds the attention of the throng with the raised index finger of the left hand. The contrasting scenes vividly illustrate the switch from the dull stasis of the exposition of inherited learning to the fervour of active discovery. Vesalius does not completely discard the ancients; he alludes to their tradition by including three prominent figures in sandals and ancient dress in the foreground, also straining to learn. The (imaginary) setting is a giant domed theatre, stone-walled, with tall Corinthian columns and a lofty gallery. It exudes a sense of moment and undoubtedly reflects Vesalius's belief in his approach and ambition for the project.

Vesalius's systematic approach and meticulous description of bodily organs were groundbreaking. Moreover, his revolutionary account of human anatomy came less than a hundred years after the invention of a revolutionary new technology – the printing press – that would transform the possibilities for depiction and dissemination. It is clear from the content, framing and presentation of *De fabrica* that Vesalius was writing for an audience. Sir William Osler (Regius Professor of Medicine at Oxford in the early twentieth century and author of *The*

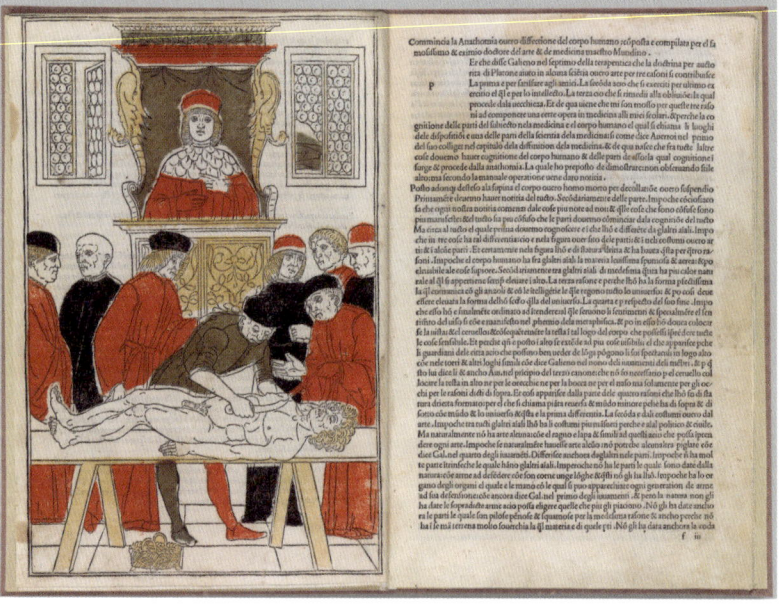

Figure 4.4 (above) Johannes de Ketham, Anatomy demonstration from *Fasciculo de Medicina*, 1494, Venice.

Figure 4.5 (opposite) Andreas Vesalius, Frontispiece from *The Epitome*, 1543, Basel.

ANDREAE VESALII
BRVXELLENSIS, SCHOLAE
medicorum Patauinæ professoris, suorum de
Humani corporis fabrica librorum
EPITOME.

CVM CAESAREAE
Maiest. Galliarum Regis, ac Senatus Veneti gra-
tia & priuilegio, ut in diplomatis eorundem continetur.

Evolution of Modern Medicine) wrote that Vesalius clearly appreciated the significance of his own work – an insightful comment and, perhaps, a tangential swipe at his professional style.[19] The only known portrait of Vesalius is found in *De fabrica*. Aged twenty-eight, he has neatly cropped hair and a full dark beard. He stands facing a dissected specimen with his head turned towards the viewer, whom he engages with a challengingly direct stare. The grandiosity of Vesalius's work, and his presence in it, could hardly be more different from the contemplative private studies of Leonardo's hand-drawn and annotated texts that were revised and refreshed over a period of years but never published in his lifetime.

Vesalius went to extraordinary trouble to make sure that the impeccable content of the work was matched by the quality of its presentation and reproduction.[20] He had woodblocks cut in Venice and sent across the Alps to Basel, as recorded in his letter to the printer Johannes Oporinus, in which he stipulated even the thickness and evenness of the paper. The question of the illustrator is debated. At times, the work has been attributed to Titian, although more probably it was the work of Jan Stefan van Calcar, whom Vesalius described as the 'outstanding artist of our age' and who had studied in Venice under Titian (*c.* 1536/7).[21] In fact, Calcar had contributed other illustrations to Vesalius's texts – e.g. *Tabulae anatomicae sex* – with surer attribution.

In any event, the quality of the woodcut prints is exceptional. The blocks later came under the control of the printer Andreas Maschenbauer, who published a selection for the use of artists. At that time, around 1706, they were presented as the work of Titian,[22] on account of the quality of the images, but had that been the case, Titian would presumably have been identified by name in *De fabrica*.

The woodblocks eventually ended up buried in the library of the University of Munich, where they were rediscovered during an

FIGURE 4.6 Andreas Vesalius, Image of the heart from *De humani corporis fabrica*, 1543, Basel. The image is hand-coloured, possibly as part of a presentation copy.

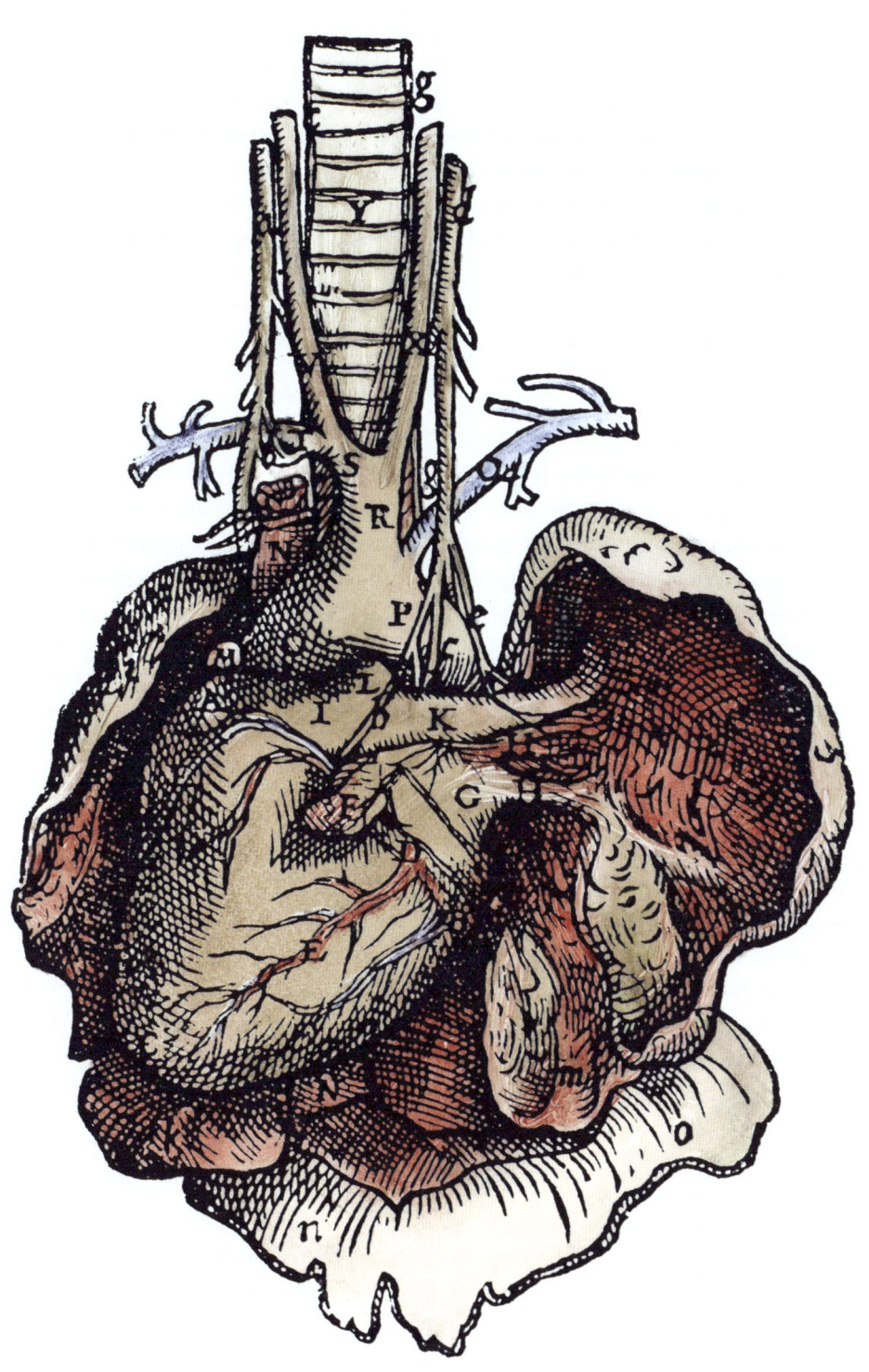

inventory in 1893. The last edition of *De fabrica* to be printed from the original woodcuts was in 1934, jointly under the auspices of the New York Academy of Medicine and the University of Munich (edited by Wiegand, Lambert and Malloch). They were eventually lost in the bombing of Munich in the Second World War.

The illustration of the heart shown in Figure 4.6 is from Vesalius's 'Epitome', the shorter version of *De fabrica* and is a hand-coloured image, possibly from a presentation copy. The depiction is qualitatively different from earlier illustrations – for example from Dryander and Eustachi. This heart is one of a number of explanted samples depicted with the surrounding tissues still attached. They are softer and show an element of realism that is completely lacking from the earlier anatomical studies. The figures are intricate. They are also cut with a certain grace that was new to the field. There is a delicacy in both the dissection and its depiction. The heart is seen from all aspects; its position in the thorax is shown, as well as its relationship with the other organs; its membranes (the pericardium) are explored in detail, along with the relationships with the veins and arteries. Obtaining this quality of dissection by removing the extraneous tissues while maintaining the integrity of the organ itself is a skilled and painstaking task.

In contrast to what had gone before, Vesalius was striving for complete anatomical accuracy, born from observation. It has been said that Vesalius was 'the consummate morphologist, totally fascinated by the details of shape and form, placement and relations, structure and consistency with an innate gift for appreciating their subtleties and blessed with the necessary skills to describe them clearly and concisely'.[23] Yet, in parts, the accompanying text was essentially an embellished restatement of Galen. In relation to the function of the heart, Vesalius wrote: 'the principal functions of the heart… consist in transforming blood for the benefit of the lung and in perfecting vital spirit'.

By contrast, Fabricius and Harvey, while benefiting from the rigour of Vesalius's approach, moved beyond flat description to ask questions about *why* the tissues and organs are structured in the ways they are. Both Harvey and Fabricius focused on function over mere structure. Harvey's approach was fundamentally modern in that it was prepared to build from first principles and to challenge dogma through observation and rational, even quantitative, analysis: 'I do not profess to learn and teach Anatomy from the Axioms of Philosophers but from Dissection and from the fabric of Nature.'

This is not the place to examine in detail Harvey's elegant and logical dismantling of the conventional beliefs, but I encourage the interested reader to consult Harvey's own text, which takes the form of a clearly constructed and incisive argument, or to refer to a number of modern commentaries.* [24, 25]

However, the key elements of his approach were to debunk what was plainly wrong (the clear absence of the septal pores, the implied rate of blood production) and to piece together an alternative synthesis. By observing the rate of blood flow from severed arteries and making some relatively crude calculations on that basis, it was possible for Harvey to show that Galen's required rate of synthesis of blood by the liver (supplied in turn by chyle from the intestine) was entirely implausible:

> I frequently and seriously bethought me, and long revolved in my mind, what might be the quantity of blood which was trans-mitted, in how short a time its passage might be effected and the like; and not finding it possible that this could be supplied by the juices of the aliment without the veins on one hand becoming drained, and the arteries on the other getting ruptured through the excessive charge of blood, unless the blood should somehow find its way from the arteries to the veins and so return to the

---

* *Harvey's Heart: The Discovery of the Blood and Circulation* by Andrew Gregory and *Circulation: William Harvey's Revolutionary Idea* by Thomas Wright are informative commentaries.

right side of the heart; I began to think whether there might not be a motion, as it were in a circle.

…Butchers are well aware of the fact and can bear witness to it for cutting the throat of an ox and so dividing the vessels of the neck, in less than a quarter of an hour they have all the vessels bloodless; the whole mass of blood has escaped.[26]

Harvey simply could not reconcile the prevailing view that blood is made by the liver and sent to the peripheries with his calculations on the rate of production of blood that would be required in such a system. He had worked out the rate of exsanguination of slaughtered animals after a major artery had been severed. The liver (and the nutrient supply required) could not possibly keep up with the implied rate of production. The powerful spurting of blood from an open artery also led him to dismiss the notion that the blood was moving in patterns of gentle ebbs and flows. So much of the existing explanation of the movement of blood did not make sense to Harvey.

Dr Jonathan Miller made the intriguing argument that Harvey's breakthrough occurred in an era of growing automation; that it was natural for him to think in terms of the heart as a pump, since mechanized pumps were coming to prominence at that time. Similarly, for Galen the reference points that shaped his thinking might have been foundries and purification processes, both of which were dependent on the production of large amounts of heat; for Aquinas, considerations of mechanics and astronomy. In other words, scientific thought in a given era borrows, naturally enough, from its broader cultural context. While this seems persuasive as a general pattern, the history of Harvey's discovery and his own comments at the time suggest that it was his conclusions about the *necessity of circulation* that cracked the riddle, rather than inferences made from the pump-like motion of the heart.

When Robert Boyle (1627–91) asked Harvey about the moment of discovery, Harvey apparently spoke of his realization of the functions of the valves in the veins. (The valves or 'little doors', *ostiola*, were first described by Fabricius in his *De Venarum Ostiolis* [*On the Valves of the Veins*, 1603].) The directionality of the flow implied by the valves – and demonstrated by Harvey – made unidirectional flow, and therefore circulation of blood, a necessary conclusion.

> Which motion we may be allowed to call circular, in the same way as Aristotle says that the air and the rain emulate the circular motion of the superior bodies; for the moist earth, warmed by the sun, evaporates; the vapours drawn upwards are condensed, and descending again in the form of rain moisten the earth again… The heart, consequently, is the beginning of life, the sun of the microcosm, even as the sun in his turn might well be designated the heart of the world; for it is the heart by whose virtue and pulse the blood is moved.[27]

Harvey was still of the view that the heart infused the blood with 'natural heat', so he would have been reassured by the concordance of his new ideas with those of Aristotle and a unifying world order.

Harvey also noted 'in opposition to commonly received opinions' that 'the arteries are filled and distended by the blood forced into them by the contraction of the ventricles; the arteries therefore are distended because they are still like sacs or bladders and not filled because they expand like bellows'. In other words, the heart is a pump.[28]

Compared with Vesalius's visually magnificent work, Harvey's seventy-four pages, almost all unillustrated, are little more than a pamphlet. In the context of considering representations of the heart, it is noteworthy that *De Motu Cordis* contains hardly any diagrams, and despite Harvey's grand statements about the primacy of the heart, its actual function is not addressed in any great detail. There are no diagrams or illustrations of its newly defined position as a pump at the hub of the circulatory system. Re-emphasizing what Harvey seems to have seen as the pivotal observation, he reserves the small number of illustrations for detailing the physiology of the patterns of filling and emptying of the veins in the forearm, highlighting the filling by the arteries at high pressure, and that venous flow is only towards the heart. The latter he demonstrates elegantly by showing how the veins fill only towards the heart, while their valves prevent backfilling from the heart.

In respect of Harvey's contribution and especially his reverence for Aristotle, historians of medicine have debated whether his approach was truly 'modern'. While he built theories from observation, they say, he also retained an Aristotelian worldview of the human as a microcosm in the macrocosm of the Universe, subject to the same laws and with a body imbued with vital spirits.[29] He even justified the circulation of blood by comparison with the cyclical nature of evaporation leading to rain. From my perspective, though, it matters little whether he aligned with Aristotle, the 'Anatomists' or the new science of the seventeenth century. In the most critical sense, Harvey was thoroughly modern. Using no new equipment, his remarkable triumph was born of the rigour of his approach. He showed a willingness to challenge the established dogma, building radical new ideas from careful observation, calculation and analysis.

His contribution was profound and rapidly, if not universally, accepted. Hobbes wrote that Harvey 'is the only man, perhaps, that ever lived to see his own doctrine established in his lifetime'.[30] Whether or not this is accurate does not matter much – clearly Hobbes had been impressed by the impact of Harvey's work, and by the time Thomas Birch published *The Heads of Illustrious Persons of Great Britain* in 1737, Harvey's reputation was securely established, such that he featured among the hundred or so plates of royalty, nobility and the likes of Sir Francis Drake and William Shakespeare. Beside the pensive Harvey, his discovery is illustrated on a board, propped above the Staff of Aesculapius and rambling poppy leaves (Figure 4.7). Out of the heart emanates a branching vascular system, while dark hatching allows the heart to stand proud of the plane of the image. It is an anatomical approximation, but the message is clear: the heart indisputably occupies the centre.

The first medical book to refer to Harvey's work as established fact was a treatise of 1651 by Nathaniel Highmore (1613–85), *Corporis humani disquisitio anatomica* (*Anatomical Analysis of the Human Body*). Besides that distinction, our interest in this book comes from its bizarre allegorical frontispiece (Figure 4.8), which also reveals the intellectual turmoil of that time.

The central figure is reminiscent of the splayed forms of the earliest anatomical depictions. On closer inspection, the figure is two-headed, a reference to Janus, the Roman god of beginnings and transitions, who presided over passages, doors, gates and endings, as well as transitional periods, such as from war to peace. Janus was usually depicted as having two faces looking in opposite directions, one towards the past and the other towards the future. Unsurprisingly, the greats of the field, Hippocrates and Galen, are shown in the principal positions. Each clasps an arm of the subject. Galen seems to be palpating the subject's radial artery pulse. Harvey (bottom right) does not achieve the same status but is nonetheless prominently acknowledged. All are placed beneath Queen Anatomy, who is enthroned with a bone as a sceptre and a skull in place of an orb.[32]

Somewhat awkwardly to our modern eye, between the legs of the subject the engraver has placed a landscape scene in which a mechanical hand pump supplies a mountain top with water that coalesces from multiple small tributaries into a larger vessel. Highmore's own annotation is more than a little flowery:

The artery, here like a cistern fixed to the Caucasus,
Expels its waterfall of blood
And rosy Maeanders to irrigate the Asia Minor of Man,
And distant provinces of the soul.
With the course having been completed up to this point,

FIGURE 4.7 Thomas Birch, Portrait of William Harvey from *The Heads of Illustrious Persons of Great Britain*, 1737, London.

The streams press out their beds into the large vein,
The dark red Danube of man
To revolve by perpetually falling,
And again to kiss the stained ocean with the purple dye of the
liver,
And to restore the vivifying heat from the pump.[32]

This neat diagrammatic device summarizes the continuous motion of the blood in circulation as the microcosm of a great macrocosmic landscape and, of course, incorporates the idea of the heart as a pump. Its connection to the minor tributaries (analogous to capillaries, which were at that time imagined but not described) rather than the larger vessels (arteries) is a minor error of detail. The dam in the bottom left of the image presumably refers to a medical ligature used in bloodletting procedures, since these would have been included in medical texts of that time, and because the curious labels *deritavio* and *revulsio* refer to specific bloodletting terminology, wherein blood was obtained respectively from the affected part of the body or from one that was remote from the ailment.[33] However, the most fascinating element is a tiny detail, almost lost. The hand on the lever of the hand pump emerges from a cloud. Whatever the appreciation of the heart and its role as pump, there was no earthly explanation for its energy or its automaticity. Life, and the energy force that propelled it, still began with God.

René Descartes (1596–1650), now better known for his contributions to mathematics and philosophy, was keenly attentive to physiology and held particularly strong views in respect of the heart. He read *De Motu Cordis* in 1632 and was convinced by Harvey's central thesis on the circulation of the blood. Interestingly from our current perspective, his dispute with Harvey's work revolved not around circulation itself, but around the question of whether the movement of the heart was an inherent property of the organ as a muscular pump (Harvey's view), or whether it was effectively a fermenting chamber or furnace wherein

THEATRVM AVTOPSIÆ

SACRVM ANATOMIÆ

Contemplationis musæum

HIPPOCRATES

GALENVS

CORPORIS
HVMANI DISQUI-
SITIO
Anatomica
IN QVA
SANGVINIS CIR-
CVLATIONEM
In quavis Corporis parte, plu-
rimis typis novis, de animal-
lium Medicorum succincta eluci-
cidatione Ornatam prosequtus est
NATHANAEL HIGHMORE
Artium et Medicinæ Doctor
nuper è Societate S.tæ Trinitatis
OXON.

Oculus Opticus

Revulsio

blood rapidly expanded its volume with increasing temperature, secondary to which the heart expanded.

> I supposed, too, that in the beginning God did not place in this body any rational soul or any other thing to serve as a vegetative or sensitive soul, but rather that he kindled in its heart one of those fires without light which I had already explained, and whose nature I understood to be no different from that of the fire which heats hay when it has been stored before it is dry, or which causes new wine to seethe when it is left to ferment from the crushed grapes.[34]

In reality, of course, core body temperature is tightly regulated and there are no such fluctuations, but the very presence of a 'core temperature' with cooler peripheries reinforced a view of the heart as a furnace and central heat source. The French physician Jean Fernel (1497–1558), who is credited with inventing the word 'physiology', had earlier undertaken an experiment which led him to declare that a finger inserted into the left ventricle of the heart of an animal finds the warmest point in the body. Descartes's theories suffer from a total lack of quantitative analysis or measurement. Even were they to be present, the changes in temperature required to effect shifts in volume would be impossibly large. That said, the connection between life, warmth and the beating heart and their opposites in death must have been irresistible.[35]

Dorothea Heitsch, a scholar in Francophone studies, writes:

> According to this theory, natural spirits are generated by food digested in the stomach and pass into the veins (this is the first fermentation in Cartesian physiology). Heat and agitation

FIGURE 4.8 Frontispiece for *Corporis humani disquisitio anatomica* by Nathaniel Highmore, 1651, The Hague.

augment these spirits in the liver and veins, allowing them to pass to the heart where heat is even greater than in the veins. The heart, through a boiling effect, rarefies the spirits (the second fermentation in Cartesian physiology) which makes the heart and arteries expand. This rarefaction separates more particles from the blood, converting them into vital spirits. These then issue forth from the great artery in the highest agitation and velocity, and go straight to the brain and through narrow conduits into its cavities where, separated from the blood, they thus become animal spirits (this could be described as the last fermentation, distillation, or rarefaction in Cartesian physiology). The spirits' passage into the brain's cavities is a mechanical process similar to screening, sieving, or filtering. At this point, the animal spirits are no longer blood, and their properties vary according to the particles that comprise them.[36]

Descartes's *The Passions of the Soul* (1649) removed sovereignty from the heart and shifted its role to one that was subordinate to the intellectual centre, the brain. In the Cartesian model, the brain was set apart from the body. In the governance structure, the heart fell below the brain, which was the central control hub. In Harvey's model, the heart is a central radiant sun – a source of heat for the body that distributes its heat via blood to the body by way of circulation. Since Descartes believed that muscles were under the control of the brain, he could not reconcile the function of the heart with the lack of conscious neural stimuli.[37] As we shall see later, this dispute is at the core of why the heart has its unique position: it is precisely the responsive autonomy of the heart that affords it its primacy among the organs.

But Harvey and Descartes approached the overarching question of physiology from very different perspectives. While Harvey still couched his observations in references to Aristotle, Descartes was striving for a new unified philosophical system that encompassed physics, mathematics, psychology, cosmology and epistemology. He viewed the body as a machine and argued that its various functions could be reduced to mechanical models, which he likened to the workings of a clock.[38]

Descartes nonetheless rejected Harvey's notion of the heart as a muscular pump. In part, this reflected Descartes's belief that muscles were necessarily controlled by the will of the brain (which, as Thomas Aquinas had noted, the heart was palpably not), and in part it was based on his misconception that the heart became hard during its relaxation phase, rather than during its contraction. Descartes was still tied up in the idea of the heart as a furnace in which blood expanded during a process of instantaneous heating (ebullition) and contracted during cooling. He supposed that there was some yeast-like property contained within the blood that remained within the heart during each cycle of contraction and relaxation.

Crucially, he saw in Harvey's work no clear explanation for the very apparent differences in the appearances of arterial and venous blood. And finally, he simply could not account for what made the heart beat:

> If we suppose that the heart beats the way Harvey describes it, we would have to imagine some faculty causes this motion, and the nature of this faculty would be much more difficult to understand than what it claims to explain.[39]

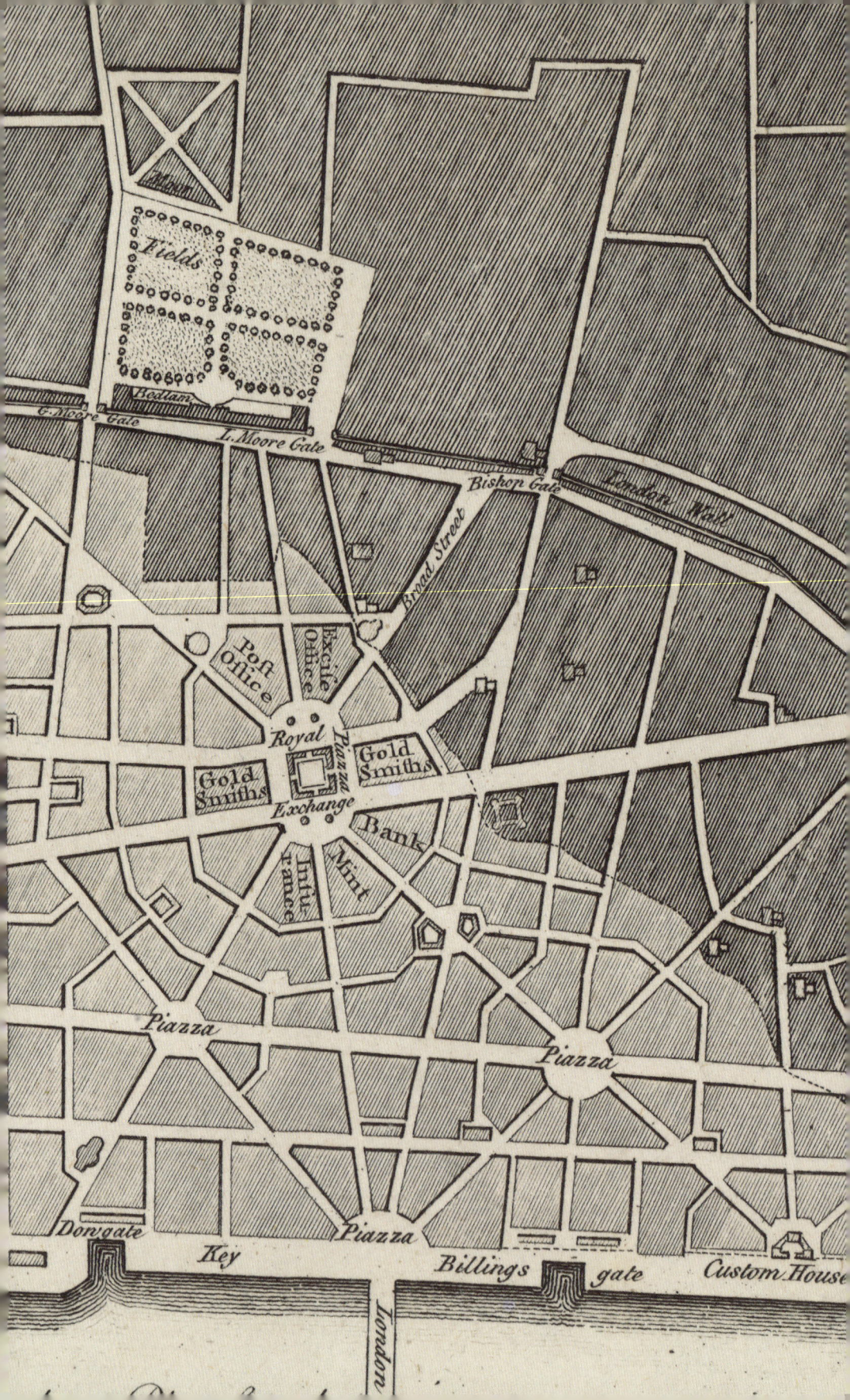

Fields

Bedlam

E. Moore Gate

L. Moore Gate

Bishop Gate

London Wall

Broad Street

Post Office

Excise Office

Royal

Piazza

Gold Smiths

Gold Smiths

Exchange

Bank

Insurance

Mint

Piazza

Piazza

Dowgate

Key

Piazza

Billings gate

Custom House

London

A Proliferation of Knowledge: Harvey and the Oxford Physiologists

*I should speak here of the ultimate way in which
the heart's movement is effected, but as it is over
difficult to obtain any true conception of this and it
is the privilege of God alone, who comprehends the
heart's secrets, to understand its movement also, I
will not waste effort in examining it further.*

RICHARD LOWER
Tractatus de Corde, 1669

Harvey's elucidation of the circulation of blood closed a long chapter of medical syntheses started by the ancient Greek physicians. At the same time, it opened up all the possibilities of modern medicine, surgery and therapeutics. Harvey did not tell the whole story, but his work instigated an extraordinarily productive period of medical research and revelation, particularly of the cardiovascular system, largely undertaken in London and Oxford.

Harvey had been appointed physician extraordinary to James I in 1618 and later to Charles I, to whom he dedicated *De Motu Cordis* (1628).[1] By then the king's 'ordinary' (day-to-day) physician, Harvey occupied a suite in the Palace of Whitehall, and at the time of the English Civil War his fortunes were closely tied up with those of the king. When Charles I abandoned his London residence, other members of the household also fled. Harvey's apartments were ransacked and, disastrously for the interests of posterity, his papers and personal possessions were destroyed or dispersed.[2]

Harvey is even said to have accompanied Charles I to the battlefield. In *Brief Lives*, Aubrey says of Harvey that Charles's sons 'the Prince and the Duke of York were committed to his care' at the Battle of Edgehill in 1642. After the battle, Charles and the Royalist forces relocated to Oxford and Harvey went with them. There, from 1643 to 1646, he was the most eminent physician of a group serving the king. In Oxford, this group of physicians and researchers, equipped with Harvey's new understanding of circulation and benefiting from a collaborative working environment, formed a chain of observers and experimentalists who were eager to understand the emerging possibilities of physiology. Among their number was Harvey himself, as well as Thomas Willis, Nathaniel Highmore, Ralph Bathurst and Richard Lower.

Harvey was made an Honorary Doctor of Medicine of Oxford in 1645 and installed as the warden of Merton College. There is little written record of his work in Oxford, though ledgers in Merton College are signed by Harvey, and it may be that his role there was mainly administrative. Aubrey also records that Harvey was involved in the study of chick embryos during the period in Oxford and that his book *Reproduction of Animals* was written there.[3, 4]

However, Harvey's reputation and discovery of the circulation of blood would already have been well known, and one imagines that his presence in Oxford must have generated some considerable excitement among the rapidly expanding scientific community. When Harvey left Oxford in June 1646, Thomas Willis (1621–75), as the leading physician-scientist, effectively became his successor. Willis had matriculated in 1636, receiving his Bachelor of Arts degree in June 1639 (he received his Master of Arts in 1642). Significantly for the unfolding of his research, the next phase of Willis's training, the study of medicine, was interrupted by the Civil War and by his service in the king's army in Oxford. Despite this, Willis was awarded the degree of Bachelor of Medicine in 1646, so it is likely that his medical studies were both interrupted and abbreviated.[5] Indeed, the awarding of his degree probably happened on the recommendation of the Regius Professor and in recognition of his loyalty to the crown, rather than for any academic distinction or accomplishment.[6, 7]

Willis's unconventional training, rather than being an historical quirk, may well have been of rather more material significance. His truncated studies resulted in his being spared some of the formal indoctrination in the established texts (Hippocrates, Aristotle, Galen) and made him more open to exploration based on his own observations. In this endeavour he would have been aided by knowledge of the methods of Harvey and of his mentor and friend William Petty (1623–87), who had graduated in medicine at the University of Leyden and been appointed Tomlins reader in anatomy.

Since 1636, statutes had made dissection part of the medical training in Oxford and Charles I had granted the reader in anatomy the right to dissect the bodies of those executed within twenty-one miles of Oxford.[8] Petty started an organized club, with regular Thursday meetings, initially at 107 High Street in the house of John Clarke, who had been Harvey's apothecary.[9] The focus of this group was the biological sciences, which distinguished it from the natural scientists meeting around the same time in Wadham College – the group that was to become the Royal Society.[10] Over time the club came to meet in Willis's house, Beam Hall, which still stands opposite Merton College.

From 1660 to 1675, Willis was the Sedleian professor of natural philosophy and a prominent figure in Oxford academic life.[11] According to the university statutes, one of the professor's duties was to read from the works of Aristotle (e.g. *De anima* and *De physica*). This pattern of transmission of accepted knowledge had been established for many years: from 1510 onwards Thomas Linacre's generosity to Merton supported medical fellows of the college to 'expound and read in the public refectory' daily 'out of the books of Galen and Hippocrates', and 'well into the 1670s the Linacre lecturers in almost all cases carried out their duties of public teaching on the classical texts'.[12] However, it seems that, having attempted to comply with the statutes in his early lectures, Willis changed tack and instead spoke about his current work, with which he would have been more familiar and, one imagines, he would have found more exciting. Willis's lectures were documented by Richard Lower and John Locke.

Locke's handwritten lecture notes (now in the Bodleian Library and spanning the period 1661–4) give an insight into the state of the group's thinking. Willis lectured on strikingly diverse topics, ranging from occasional and habitual melancholia, sleepwalking and epilepsy to colic, flatulence and laughter. But his cardiocentric classification of the forms of animate life are worth a brief mention:

> Various are living creatures can be divided into four classes:
> 1st and lowest: those whose life consists of the rather slow, viscous, rhythmic flow, containing within itself, the spirit activating them. These include plants, worms and other similar creatures whose parts are, as it were, homogeneous.

2nd: those, such as insects, who even if bloodless yet possess something analogous to blood, circulating throughout the body along pulsating vessels which also convey sensory function.

3rd: those who, though they have blood, yet like a second cardiac ventricle, and hence their blood does not become so heated as to need ventilation in the lungs.

4th: the most highly developed animals, you need hot blood for the regular functioning of their senses and other operations. Their hearts possess twin ventricles, and blood flows from the right ventricle to the left, through the lungs, where it is tempered by the air breathed in. The lungs also ventilate the heat of the blood; discharge its waste matter and, as some believe, take some nitrous element from the air necessary for generating spirits. For, while the embryo is confined within its mother's womb it makes no use of its lungs nor does the circulating blood, as in that state, it is devoid of movement and is insufficiently heated to the degree necessary for confirming movements. But as soon as is drawn in, and the vital flame gets the opportunity to expand, then the heat becomes much more intense, and hence the need for continuous respiration. Our life, therefore is very similar to that of the burning lamp, and the man lives as long as his blood is being burnt in his heart.[13]

The analogy with the vital flame is a little wide of the mark but again reflects the nature and limitations of contemporary understanding of the workings of the world, and the tendency of the scientists of the era to apply a flawed model to the question at hand. The blood itself is not 'burnt', and certainly not in the heart. But how could Willis have anticipated the future knowledge that oxygen is indeed carried in the blood in molecular form, attached to haemoglobin, to be consumed by mitochondria (in a process termed 'respiration') and with the generation of both heat and carbon dioxide and the liberation of vital energy? Therefore, the notion of life as 'a burning lamp' is perhaps not so very inaccurate. Willis would have been assisted by the knowledge that arterial blood is bright red because of the oxygen it carries, while venous blood, depleted of oxygen, is dark brownish-red.

After the Civil War and Harvey's departure, the Oxford scientists devoted themselves to expanding their knowledge in the modern way, through observation, dissection and experimentation. A group that included (a young) Christopher Wren, Thomas Willis and Richard Lower undertaking cadaveric dissection began to focus on the blood supply of the brain. Using a technically novel approach, they extracted the brain, whole and intact, allowing for meticulous study of its surface – specifically the underside, from where the blood supply originates.[14, 15] Wren developed the technique of *chirugia infusoria*, in which coloured liquids were injected into blood vessels to map their course. This approach was adopted by Boyle, Willis and Lower. As early as 1651, Lower informed Boyle that:

> The Doctor [Willis] does likewise intend… to syringe in some kind of liquors tinctured with saffron or other colours into the arteriae carotids, the brain being first opened, just after the creature is dead, and warm, to try how the blood moves, and how the tinctures may be separated in the brain.

As part of these investigations, Willis and Lower – using a dog for experimentation – injected *one* of the two carotid arteries that supply the brain with blood. The ink they introduced into the arteries illuminated a network of capillaries covering the surface of the *entire* brain. The fact that injecting the artery supplying one side of the brain distributed blood to the surface of the brain on both sides was highly significant. On closer examination, it became apparent that two arteries on each side of the brain (i.e. four in all) joined together in a circuit at the base of the brain, from which arose branches that penetrated the brain itself.

The blood supply to the brain had evidently evolved such that in the event of an interruption to the main supply on one side, the supply from the other side could compensate. The circulatory connection that links the blood vessels and allows this to happen is known as the 'circle of Willis', in honour of its discoverer. Willis identified not only the structure but also its function. He recollected a dissection he had performed in which one carotid artery was entirely occluded (with what one assumes was atherosclerotic material):

'For the carotidal and vertebral arteries have so many anastomoses, so divinely contrived inside the dura mater,' stated Lower '… before they go up into the brain… that if three arteries were quite obstructed the fourth would convey blood into all parts of the brain and cerebellum and sufficient enough for life and motion.'[16]

To test their deduction, Willis tied off the carotid artery of his spaniel. Rather than suffering a catastrophic stroke, the dog made an immediate and full recovery. With the ligature still in place, the interconnectivity provided by the circle of Willis maintained sufficient blood supply to nourish the entire brain. The wider scientific world, already familiar with Harvey's work, soon accepted Willis's findings. Those findings were, of course, entirely accurate and have had important consequences for neurosurgery and stroke medicine, in which understanding the particulars of the blood supply to the brain (and the tolerance to potential interruption) is crucial.

By 1663, the experimentation was complete and Willis was preparing for publication of what was to become the first textbook in neurology, *Cerebri Anatome*. Richard Lower undertook much of the dissection and drew the neuronal structures, while Christopher Wren depicted the vascular structures. A noteworthy aspect of this period of discovery and experimentation is the remarkable versatility of certain of its practitioners, including Robert Hooke, Christopher Wren and Robert Boyle. Such men were not constrained by the boundaries of academic or professional disciplines and seemed to have made ample use of the chance to move where their interest and opportunity took them.

A case in point is that while Wren was completing the anatomical drawings for Willis's book, he also received his first major architectural commission. Having completed the design for the chapel (now the Old Library) at Pembroke College, Cambridge, Wren received an instruction from Gilbert Sheldon, chancellor of Oxford University, to design a theatre in Oxford. The Sheldonian Theatre, as it came to be known, was built between 1664 and 1669 and is one of the most distinctive structures in the centre of the city; it remains the venue for the university's major ceremonial events.

Perhaps less well known than the intellectual breadth of Wren and his peers are those instances where there was spillover from one of their fields of activity to another. Then, as now, 'design' in the natural world could be reworked and applied to inform human inventiveness. Wren provides us with a striking example.

Following the Great Fire of London in 1666, King Charles II called for proposals for rebuilding and reconfiguring the layout of the city. Plans were submitted by, among others, John Evelyn, Robert Hooke and Christopher Wren. None were realized, because reparcelling the land would have been too slow and costly, and the rebuilding surged on without the realization of any unified vision, but Wren's design is of particular interest. In his book *The Craftsman*, Richard Sennett noted that Wren drew on the notion that roads are like arteries and veins and for the first time developed the idea of the one-way street (echoing Harvey's valves).[17] But comparison of Wren's proposal for London with his illustration in *Cerebri Anatome* shows something perhaps even more intriguing. Wren had proposed a system of interlinking roads, joined at nodes with multiple inflows. In so doing, he seems to have been drawing not only on Harvey's principles of (one-way) circulation of the blood, but also on his own work with Willis – specifically the multiplicity of linked feeding arteries found in the brain. Wren was employing principles of physiology and the structures he had depicted in Willis's book to inform a rational design for the rebuilding of London. If one superimposes one-way flows of traffic, the analogy pulls us even closer to the idea that Wren used his newly found knowledge of the circle of Willis to design a street layout that would maintain efficient traffic flows despite blockages or interruptions to any one access (Figures 5.1 and 5.2). This minor digression illustrates vividly the evolution from descriptive anatomy to experimental physiology, the rapidity of transmission of new ideas and the highly collaborative environment that existed in England in the second half of the seventeenth century.

After Willis and Wren, the third member of this exceptional Oxford group was Richard Lower (1631–91), who was arguably the first cardiologist and would become the most prominent physician in London of his time.[18] Although his contributions are probably less well known, or at least less readily attributed to him, they added crucial new knowledge of the structure of the heart muscle and its autonomous function.

Lower acknowledged that the heart had been known as a muscular organ since the time of Hippocrates and that Harvey had noted its tension, contraction and hardening. But linking this to purposeful function would be one of Lower's key contributions. To the extent that the heart is a pump, it is not merely a rigid mechanical piston. It is highly engineered, with a complex three-dimensional architecture that determines its wringing motion. It is also exquisitely sensitive to moment-to-moment physiological variations and can adapt its force of contraction to adjust for changes in the volume of blood it accepts – for instance, during exercise.

Lower dismissed the lingering notion that the cyclical expansion of the heart was due to 'ebullition' or 'ferment' of the blood. Instead, he concentrated on meticulous study of the heart's composition and structure to tease out how it might function.

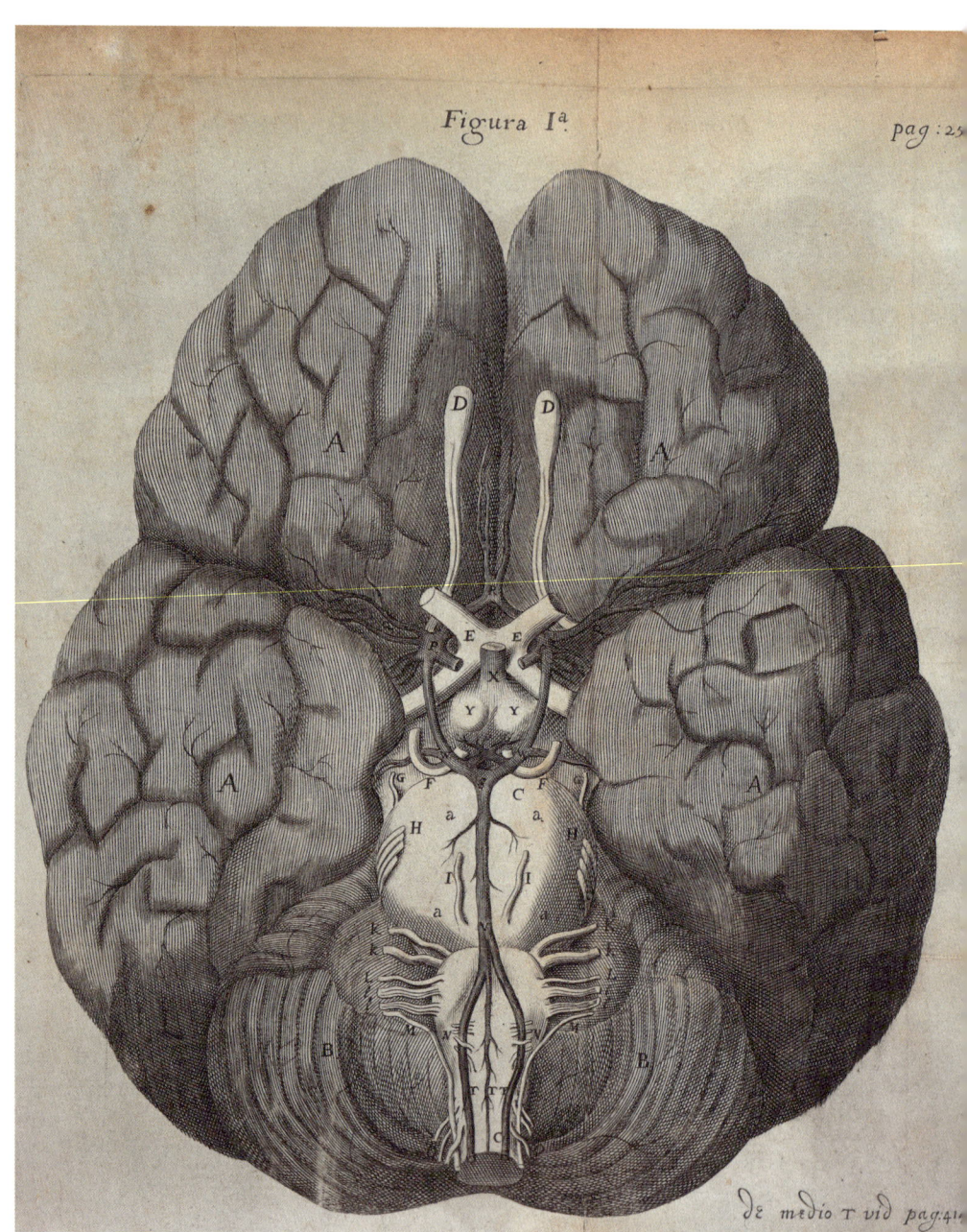

de medio T vid pag:41

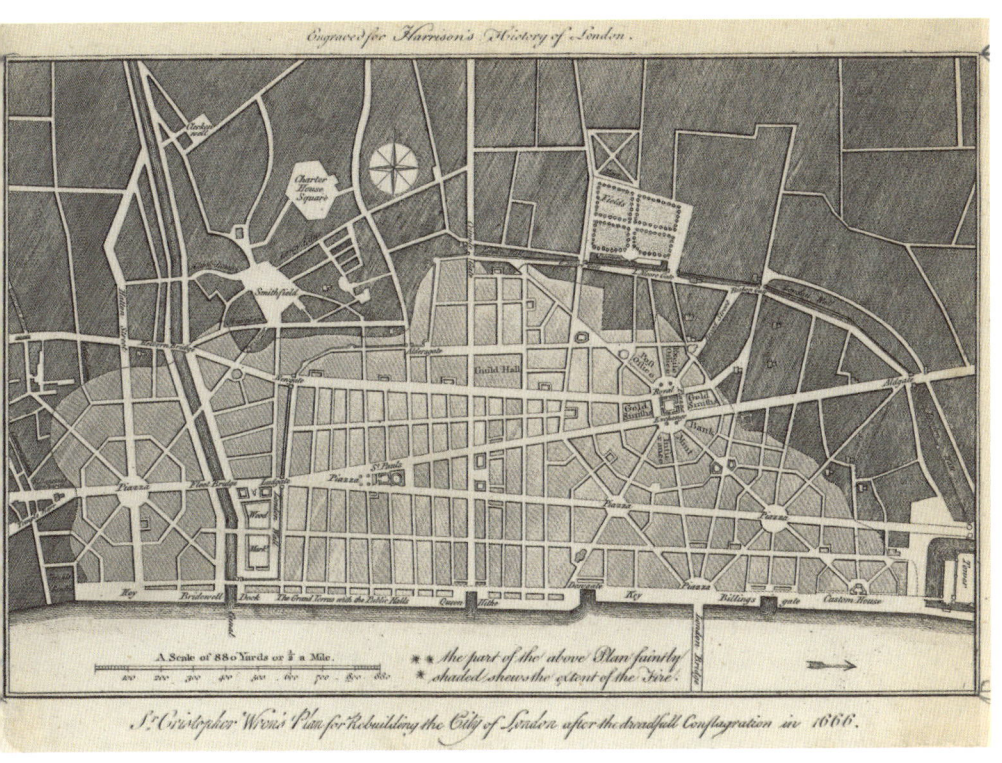

FIGURE 5.1 (opposite) The arteries in the base of the brain, seen from below, in Thomas Willis, *Cerebri Anatome*, 1664, London.

FIGURE 5.2 (above) Sir Christopher Wren's road design for the rebuild of the City of London after the Great Fire, *c.* 1666, London.

In a heart that has been well boiled and had its auricles and larger vessels removed, there is seen a fairly strong tendon which passes right round and encircles the edge of the heart about its openings… The fleshy fibres which enfold and make the external surface of the heart pass upwards and to the right everywhere to be inserted into this tendon. The inner fleshy fibres on the other hand, which lie next to the ventricular cavities, are inserted into the same tendon in exactly the opposite direction.

Now that it is clear that the Heart's fibres end in two different ways it is necessary to show that they also encircle the whole circumference of each ventricle in a similar sequence with the exception of a few rather delicate fibres, which are carried straight up over the external surface of the right ventricle.

…the fibres immediately underneath these straight external fibres in the right ventricle pass up obliquely to the right to terminate in the base of the Heart, and by their spiral course recall quite well Helix or the snail.

…the orderly sequence and infolding of all these fibres will easily be grasped by anyone who tries dissecting the Heart of an ox or a sheep.[19]

In fact, what Lower is describing is the helical orientation of the muscular fibres in the heart. In his *Tractatus de Corde* (*Treatise on the Heart*) he illustrated this meticulously (Figure 5.4), showing that the fibres are not oriented in concentric loops that simply encircle the ventricular cavity, but that they form spirals.[20] This was confirmed by Dr Patrick Hales and colleagues using sophisticated *in vivo* magnetic resonance 'diffusion tensor' imaging techniques that defined the route and orientation of the fibres by measuring the diffusion of water along their paths (Figure 5.3).[21] As a consequence of this three-dimensional configuration, when the fibres contract, the heart expels the blood through a mechanically efficient wringing motion. This elegant 'design' also contributes to the highly coordinated patterns of blood flow through the heart on each beat. Using magnetic resonance velocity mapping, Professor Philip Kilner and colleagues described (appropriately poetically) the 'sinuous, chirally asymmetric paths of flow' (Figure 5.5)[22] that pertain in the right-sided chambers (receiving blood from the peripheries and propelling it to the lungs) and left-sided chambers (receiving blood from the lungs and distributing it outwards through the aorta). On the basis of the mapped flow patterns, they proposed potential fluidic and dynamic advantages in which the momentum of inflowing streams of blood leads to optimized flow, minimizing the dissipative churning between entering, recirculating and outflowing streams.

If Harvey elucidated the fact of the circulation of the blood, it was Lower who promoted the role of the heart, and the nature and structure of the muscle fibres that effect its pumping action and the sequential contraction of the atria and ventricles. He identified the combined roles of the 'restitution' of the ventricles and the contraction of the atria, in effect the filling of the ventricles in diastole (the relaxation phase). He recognized the distinction between coronary circulation, which nourishes the heart muscle itself, and the 'impervious' nature of the ventricular cavity, which insulates the blood from its pump.

Finally, in what I think was his most important contribution, Lower tackled, head on and through experiment, the critical question that had trickled down from antiquity. He sought to determine whether the movement of the heart was a consequence of the heating of the blood (due to a putative fermentation and expansion, termed 'ebullition'), or whether the heart was the originator of its own motion. Descartes had been a proponent of the former notion, believing that the heart was not a pump but a furnace. He posited that the heart heated small particles in the blood, causing them to expand or 'rarefy'. The dynamic expansion of the blood within the heart caused the organ to swell. When the heated blood was released into the arteries, it caused them to expand in turn. Descartes reasoned that as the blood cooled it took up less space, and the heart and vessels collapsed.[23, 24] Lower put the issue to rest:

> Finally, the movement of the Heart is shown to be independent
> of any ebullition of blood by the fact that a Heart taken from
> a living animal and entirely emptied of blood does not cease to
> move, even if it is cut into small pieces.
> But to decide experimentally whether or not any ebullition of
> blood helped the blood's movement at all, it occurred to me to
> see if the heart would continue its movement undiminished after
> I had drawn off all the blood and had replaced it intravenously
> by an equal quantity of other fluids less likely to become lighter
> or to froth up. With this object in mind I drew off through the
> jugular vein of a dog almost half of its total blood injecting instead
> through the crural vein an equal amount of beer mixed with a
> little wine. This procedure I repeated several times in succession
> until instead of blood the fluid coming from the vein was merely
> a solution with less colour than the washings of meat, or than
> claret several times diluted. The Heart meanwhile became only
> slowly more feeble, so that practically the whole of the blood
> was replaced by beer before life was replaced by death.[25]

Lower's experiment was as flamboyant as it was inhumane – certainly
by today's standards. Nonetheless, it was undeniably effective. He had
shown the autonomous action of the heart muscle, its function and

FIGURE 5.3 (opposite top) Contemporary diffusion tensor imaging reveals
spirally arranged muscle fibres, 2013, Oxford.

FIGURE 5.4 (opposite bottom) Richard Lower, illustration of the spiral
orientation of the heart muscle fibres depicted in *Tractatus de Corde*, 1669
(this version, 1686), London.

FIGURE 5.5 (overleaf) The 'sinuous, chirally asymmetric paths of flow' described
by Kilner and colleagues using 4D MRI, 2000, London.

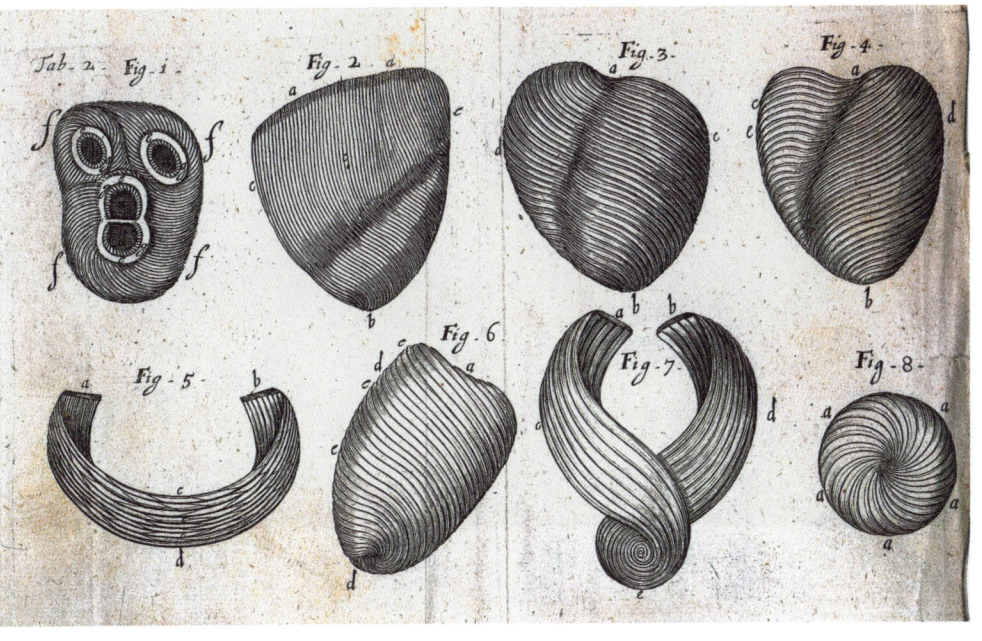

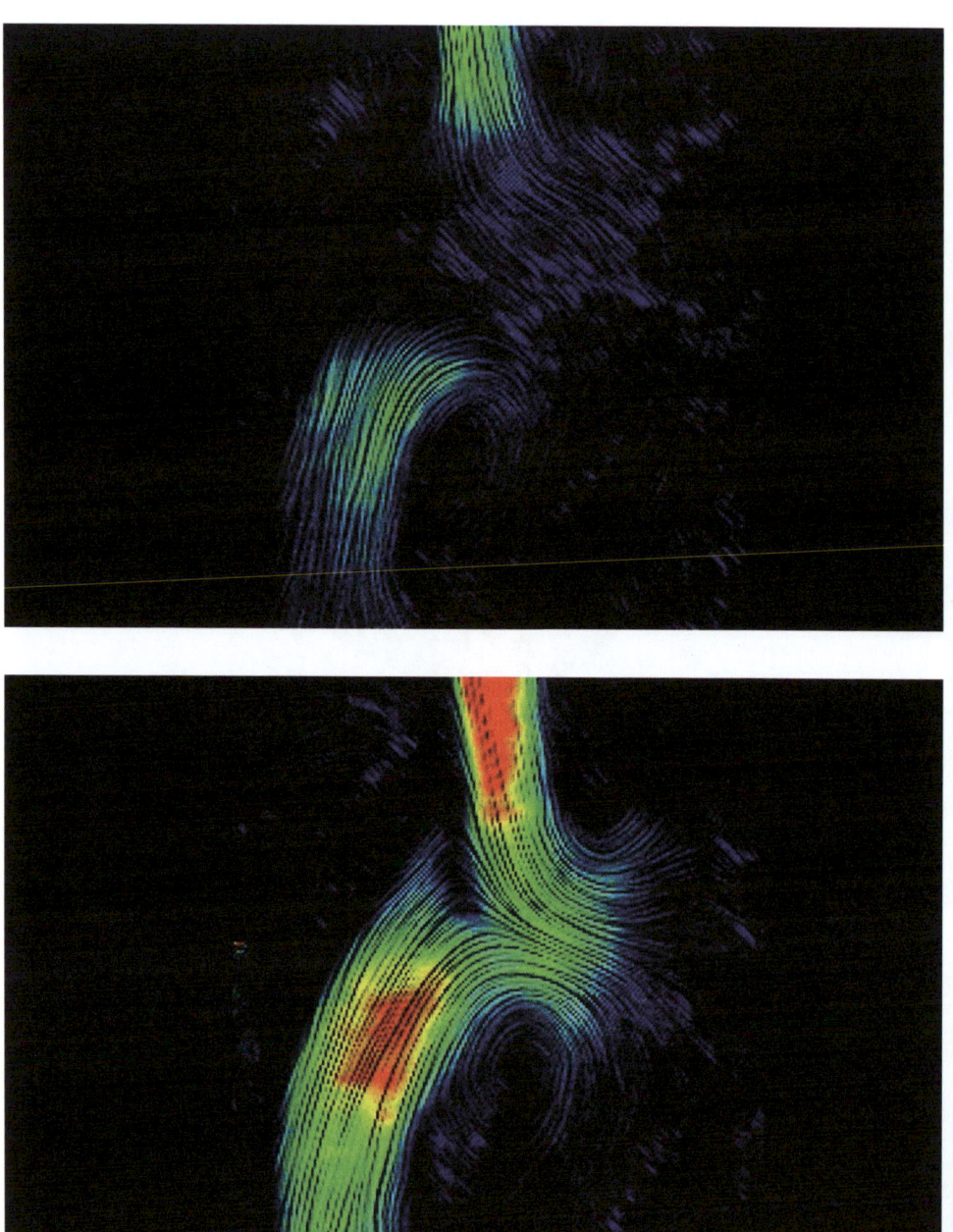

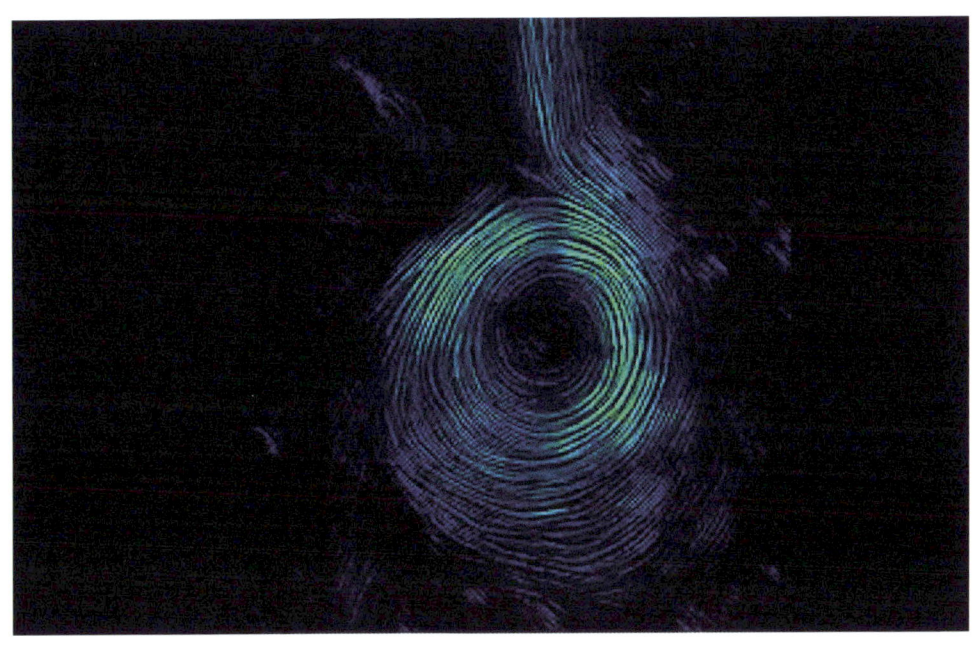

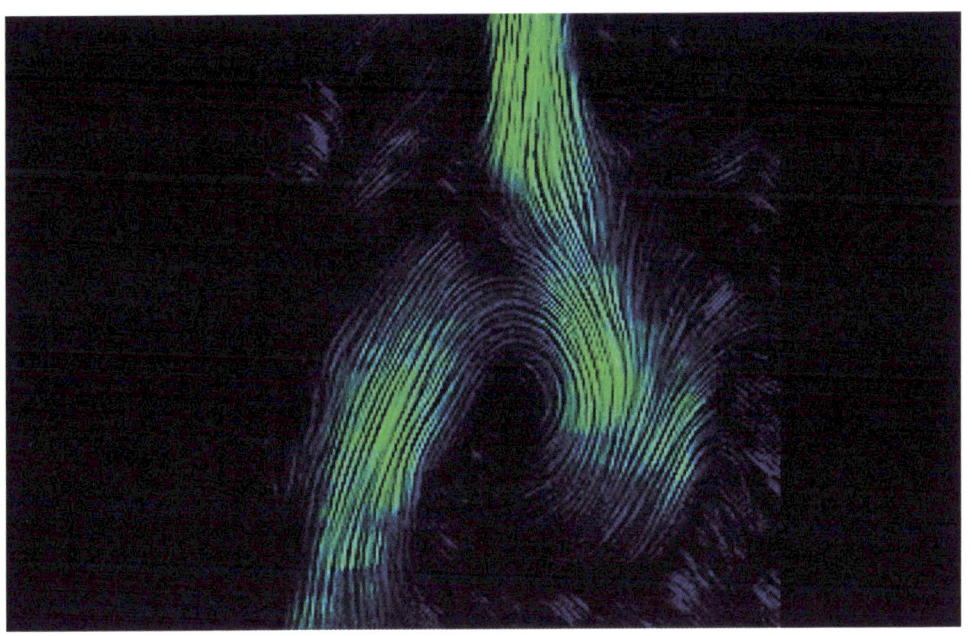

the delineation of the cardiac cycle, dispelling long-held views on the heart as a fermenter or furnace, effectively bringing together the modern understanding of the heart. However, while he brought into very clear focus the actual working of the heart as an effector of blood pumping, he could not progress further than that, and neither did he have anything like the tools required to do so. He seems to have recognized the next logical question and his utter inability to approach it. His plaintive comment brings to mind the hand of God reaching down to operate the pump lever in the Highmore frontispiece (Figure 4.8):

> I should speak here of the ultimate way in which the heart's movement is effected, but as it is over difficult to obtain any true conception of this and it is the privilege of God alone, who comprehends the heart's secrets, to understand its movement also, I will not waste effort in examining it further.

Having demonstrated that the heart has autonomous function, certainly independently of the blood and even – at least transiently – when isolated from the body, Lower has no explanation for how this came about. He made observations about the innervation of the heart, and from this (somewhat illogically, or at least inconsistently, given some of his earlier statements about its autonomy) he goes on to state with a tone of some certainty that the brain controls the heart*:

> on the mutual service rendered by brain to heart and heart to brain sensation and movement essentially depend.

---

* Lower might well have been unable to reconcile his own findings. He knew the heart was a muscle, and the muscles were usually controlled by nerves that came from the brain via the spinal cord. In addition, he identified the innervation of the heart. On the other hand, he had also shown that the heart could beat of its own accord (without attached nerves). These apparently contradictory findings were not wholly resolved until the twentieth century (see Chapter 8).

By the heart's movement blood is continuously sent to the brain and cerebellum for the production of spirits. The sprits, in turn, flow into the heart through the nerves and ensure it is in perpetual and constant movement.

Lower's analysis must have seemed perfectly plausible – after all, were not all muscles controlled by the brain? Moreover, this interpretation fit with the studies of the nervous system and drawings of the innervation of the heart that he had already provided for Willis's book. His schematic diagram, reproduced in Figure 5.6, shows an array of nerve fibres extending to the heart. Had he been entirely right that the heart was subordinate to the brain, the heart would have been rendered a diminished organ. But Lower had shown that the heart could beat in isolation. The combination of its fundamental autonomy with physiological responsiveness is what makes the heart pre-eminent. In fact, the innervation of the heart that Lower had identified was described in Willis's book. Lower had contributed to the dissection and depiction of the autonomic nervous system – that is, the nerves that extend beyond the brain and spinal cord to influence involuntary functions (such as heart rate control), though Lower and his contemporaries did not appreciate that function.

Willis was generous in his acknowledgement:

And here I made of the Labours of the most Learned Physician and highly skilful Anatomist, Doctor Richard Lower, for my help and Companion; the edge of whose Knife and Wit I willingly acknowledge to have been a help to me for the better searching out both the frame and offices of before hidden Bodies.[26]

For now, we should note that the autonomic nervous system links emotion and physiological responsiveness to the heart. Much of what we experience of the heart relates to changes in its rate and force of contraction, which are mediated by this component of the nervous system. We will come back to it later, since the elucidation of how this happens had to wait until the twentieth century, but knowledge of the *fact* of the innervation of the heart came from Lower and Willis. This interplay between the responsiveness of the heart and our awareness of it above other organs, so-called interoception (i.e. the perception of internal visceral signals), underlies our perception of the heart and the properties that we therefore ascribe to it.

In considering how the heart has been depicted, I have suggested previously that its colour, or more precisely the colour of the blood contained within it, has been influential. If one examines the blood returning to the heart from the body through the great veins, it is dark red. Conversely, the blood emerging at high pressure into the aorta is bright red. At some level, the belief that the heart heated the blood and invested it with some vital property was perfectly reasonable to early vivisectionists. An *almost* careful observer could easily conclude that the heart was responsible for the transformation of the blood; Descartes would have fallen into this way of thinking. But more careful observers appreciated that the blood exiting the heart through the pulmonary artery exclusively *to the lungs* was dark, but when it returned to the heart *from the lungs* it was bright red. In other words, the change in quality of the blood occurred in the lungs, not the heart. Indeed, a particular historic challenge to the notion of the circulation of a blood pool had been the demonstrable difference between venous (deep red) blood and arterial (scarlet red) blood: given the evidence for two types of blood, how could there be a single common circulation?

Servetus, the Spanish humanist and polymath (*c.* 1509–53) wrote of the passage of the blood through the lungs and of the change in the colour of blood that accompanied that transit, though copies of this text were burned as heretical and his views were not widely known

FIGURE 5.6 Illustration of the nerve supply to the heart, depicted by Richard Lower in Thomas Willis's *Cerebri Anatome*, 1664, London.

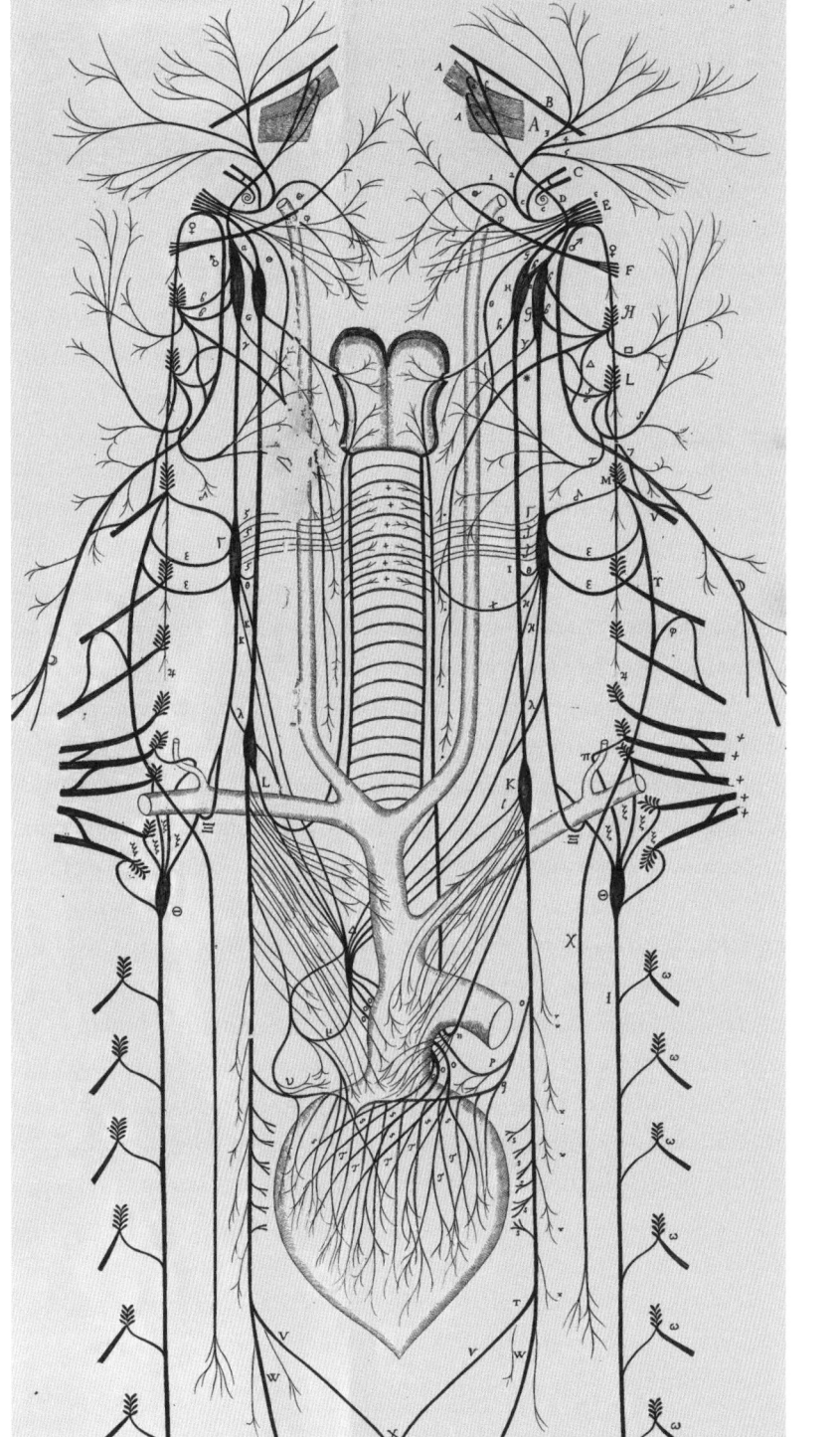

in his lifetime. Similarly Realdo Colombo (*c.* 1515–59), professor of anatomy and surgery in Padua, reported the transit of blood through the lungs and demonstrated that the pulmonary vein (carrying blood from the lungs to the heart) contained blood rather than air (as had previously been asserted), and moreover that this blood was bright red, like arterial blood. Although I will not consider the matter in detail here, the Oxford physiologists – notably Lower again, and John Mayow – also concerned themselves with the question of the change in the colour of blood, the alteration in its composition that such a change implied and the vital component of air that led to its observed effect on blood as it traversed the lungs.

There remains one last component of the circulatory system that I need to describe briefly in order to complete the story. The blood moves from the arteries into the tissues and from the tissues into the veins. The clear implication was some form of microvasculature. The tissue-level small blood vessels (which Leonardo knew must exist but could not identify) were identified by the Italian physician Marcello Malpighi (1628–94), professor of anatomy at Bologna, Pisa and Messina. He used a microscope to examine and describe the histology of the lungs, kidneys, spleen and liver.[27] This led to his discovery, in 1661, of capillaries, which completed the route for the circulation of blood. The next major advance in the scientific understanding of the heart, however, would have to wait for 300 years.

The period from *c.* 1540 to 1660 had seen a radical intellectual revolution. The grip of long-entrenched classical dogma on the heart and blood had been shattered by a combination of intellectual rigour and intensive enquiry and realized by sustained international collaboration. For the heart, the headline problems were largely resolved. Harvey had shown that the blood circulated, pumped by the heart. Lower had shown conclusively that the heart was not only automatic, but that its function was not dependent on the properties of the blood within. And finally, the cumulative efforts of several investigators had shown how air entering the lungs was responsible for changes in the composition of blood, which we now recognize as oxygenation. Ideas of spirits and vivification were banished. The circulation was a closed dual-loop system with the heart at its centre.

In some respects, the puzzle was solved. The mysterious beating spot that heralded life on the surface of the chick embryo – the source of heat and motion and semen, the seat of the soul and the recorder of deeds – was, in fact, a pump.

The Immaculate Heart

Le cœur a ses raisons que la raison
ne connaît point.
('The heart has its reasons that reason
does not know.')

BLAISE PASCAL
Pensées, 1669

The embalmed severed head of Antonio Scarpa rests in an oak-edged glass box, lodged in a high niche in the Museo per la Storia dell'Università di Pavia. We do not have a precise account of how it came to be there, but Scarpa, professor of anatomy at the university, was famously detested. The story goes that, after his death, his own assistant mercilessly dissected the body, preserving his fingers and urinary tract (he had died from an infected kidney stone) for public display. The embalmed head only came to light later. The gruesome tale, and the gory sight of Scarpa's whiskered head squinting through the formalin, his wrinkled, sunken cheeks slumping into a toothless death grimace, contrast starkly with the serene beauty of his *oeuvre*.

Having been tutored by his uncle, Canon Paolo Scarpa, Antonio entered the University of Padua in 1766. There, he was taught by Giovanni Morgagni, generally acknowledged as the founder of pathologic anatomy. Although Morgagni died shortly after Scarpa received his degree in 1770, Morgagni helped him obtain his first academic post at the age of only twenty (professor of anatomy and clinical surgery at the University of Modena). Soon afterwards, he was also appointed chief surgeon at the city's military hospital.[1] In 1781, the Duke of Modena, aiming to improve medical education, sent Scarpa to work in Paris and London. There, he studied under John and William Hunter and attended their lectures. Whether Scarpa attended William Hunter's lecture on the 'discovery' of the Windsor Leonardo drawings (see Chapter 3), I don't know, but it is plausible given the timing, and he must surely have shared Hunter's knowledge of their existence and significance; Hunter died in 1783, the year that Scarpa returned to Modena.

Shortly after his return, he was offered the Chair of Anatomy at the University of Pavia, where his research activities were boosted by an order of the modernizing and reforming Habsburg emperor Joseph II, which allowed patients who had died in the hospital at Pavia to be dissected.[2] Scarpa did not disappoint. In his *History and Biography of Anatomic Illustration* (1852), Choulant describes Scarpa as 'one of the most excellent men of his day, inventive and of untiring diligence'. Those characteristics, perhaps further inspired by knowledge or sight of the Leonardo drawings, sit well with what we see of Scarpa's work.

His most distinguishing contribution was to elevate the printed depiction of detailed anatomical structures, especially of the nerves, to a new level. He had himself trained as an engraver, and it is hard to disagree with Choulant's analysis that: 'His anatomic prints are… models of anatomic representation as regards faithful differentiation of the tissues, correctness of form and the utmost perfection of engraving.'[3] In 1794, he published his masterpiece: *Tabulae neurologicae ad illustrandam historiam anatomicam cardiacorum nervorum* (*Neurological Plates to Illustrate the Anatomical History of the Cardiac Nerves*).

The magnificent life-size figures were drawn by Scarpa himself and engraved by Faustino Anderloni. The delicately hatched life-size engravings that they created are arguably the high point of anatomical depiction. Seen in the flesh, one is struck by their size: they are presented in large folio format that adds impact to images that are already palpably realistic. The lines are crisp, and the dynamic range of the greyscale gives the images immense texture and depth. From not very many paces, they are almost photographic (Figure 6.1). The ability to convey this level of detailed precision enabled the major focus of Scarpa's work on the heart. He was interested in its supply of nerves, which he shows as wispy white strands traversing the surface of the muscle (Figures 6.1 and 6.2). It has been said that Scarpa was the first to demonstrate how the nerves to the heart branch and come into direct connection with the muscle fibres of the heart. Others had shown that the blood vessels of the heart are accompanied by nerves, but Scarpa has been credited with the discovery that cardiac muscle itself is supplied with nerves.[4] I am not sure that this is right. While he may not have described them with such visual precision, Richard Lower

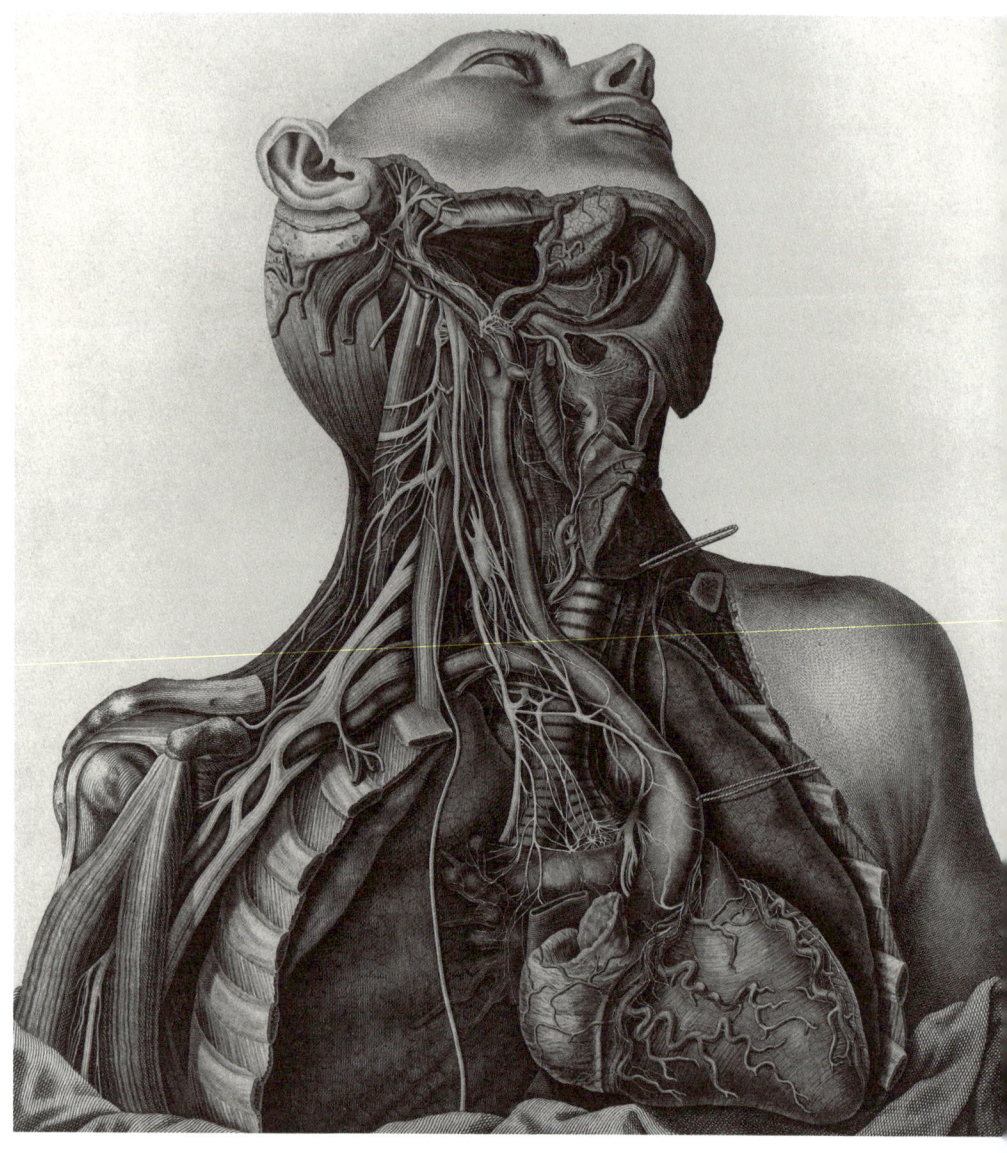

FIGURE 6.1 (above) Antonio Scarpa, *Tabulae neurologicae ad illustrandam historiam anatomicam cardiacorum nervorum* (*Neurological plates to illustrate the anatomical history of the cardiac nerves*), 1794, Pavia.

FIGURE 6.2 (opposite) Detail from the engraving above showing the nerves on the surface of the heart as thin white structures.

160

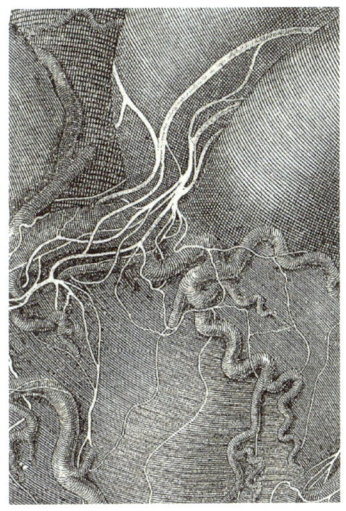

seems to have made the fundamental discovery about a century earlier:

> On the subject of the nerves which are embedded in the heart earlier authors, who were ignorant of the movement of the heart and blood, were mostly silent, and not unnaturally. Next after them come those who recognise, it is true, a circulation, but believe it to proceed so slowly and in so tortoise-like a fashion that they say blood is poured out drop by drop and only leaves the heart when it bubbles over; little concerned therefore whether or not the heart's movement helps the circulation of the blood, they attach little or no importance to the muscular structure of the heart and its numerous nerves. If, however, one considers the tendinous and fibrous material of which the heart is made and how it is interwoven everywhere with so many nerves, one must conclude that this endowment was not without purpose, but that it was constructed and set up to fulfil the same function as do the remaining muscles.
> It receives many nerve-fibres and offshoots from the nerves of the eighth pair, all of which give of various branches to each auricle, as they pass along between the pulmonary artery and the aorta and are then distributed widely to the heart substance.[5]

What Lower is describing and Scarpa and Anderloni are depicting is the relationship between the eighth cranial nerve, known as the vagus nerve, and the heart (Figures 6.1 and 6.2). A complementary set of nerves arises from the so-called sympathetic ganglia, small nodes linked to the spinal cord. The anatomical descriptions are dry but, of course, the discovery of functional connectivity of brain and heart is critical and we will return to it later.

Even as the anatomists were refining the naturalistic renderings of the heart, epitomized by Scarpa and Anderloni, whose dedication brought the art of anatomical engraving to a level of near perfection, a parallel (and eventually globally pervasive) pattern of depiction was emerging in Catholic religious art. Surprisingly, at least to me, these two strands of depiction – the hearts of the anatomists and the Sacred Hearts of Catholic devotees – were to become intimately connected.

Of the thousands upon thousands of images of the Sacred Heart, let us start with those by the Mexican artist Juan Patricio Morlete Ruiz (c. 1759) (Figures 6.3 and 6.4). They are among the most visually florid of such depictions and are rich in iconography. The first of these two allegorical works shows the Sacred Heart of Jesus accompanied by the Immaculate Heart of the Virgin Mary.

These oil on copper paintings are each dominated by the image of a bleeding heart, floating in a celestial scene of indigo. The first is encircled by a crown of thorns and has a cross inserted into a vessel that emerges from its base. The heart is pushed forwards by the bright light that seems to emanate from within. Through a hole punched in its wall, we can peer into the dazzlingly illuminated chambers. In fact, the light is so bright that the image of the crucified Christ at its centre is almost obscured. Tonally linked to the heart is the figure of God presiding over the whole, his hand raised in blessing. The heart is flanked by cherubim and beneath it sits the Virgin Mary, accompanied by St John and Mary Magdalene, who are comforting her. The intense pain of her loss is signified by a sword through Mary's heart, which is further amplified in the companion image. The figures in the medallions on each side are Jesuit devotees of the Sacred Heart, St Ignatius of Loyola (1491–1556), the founder of the Jesuit Order, St Luis Gonzaga (1568–91),

St Philip Neri (1515–95) and St Francis Xavier (1506–52). Prominent historic figures of the church, Bishop Augustine of Hippo (354–430 CE), a Franciscan friar likely to be Bernard of Clairvaux (1090–1153) and Francis of Assisi (1181–1226), accompany a woman dressed in white, representing a purified soul. All gaze at the heart.

The second image is dominated by the radiant Immaculate Heart of Mary, pierced by a sword, symbolizing her profound grief at the death of her son. The flames surrounding her heart allude to love and are more intense than those in the heart of Jesus. Lilies sprout from vessels at the top, denoting purity and perhaps symbolic of resurrection or Christian renewal. The dove of the Holy Spirit hovers over Mary's heart, while Jesus himself sits beneath, cradling his own heart in one hand and accompanied by St Joachim and St Anne. The eight medallions at the edges of this scene are framed by volutes and contain images representing the saints from the main female religious orders, such as St Catherine of Siena, St Gertrude, St Theresa of Ávila, St Rosalia and St Rose of Lima.[6]

Besides the many allusions to Catholic tradition, there is much here that resonates with heart-related tropes from very different traditions. The central spout from which the flowers emerge in the second image is reminiscent of the Egyptian representation of the heart as both a vase or vessel and an originator of life. The spout and the adjacent large vessel are borrowed from the anatomical depictions, incorrect in their precise form but sculpted to fit the artist's allegorical purpose. The crucifix within Christ's heart in the first image takes us back to Santa Chiara, and, in parallel, to Rama and Sita in the heart of the Hindu god Hanuman, while the heart as a source of heat and light draws from the

FIGURE 6.3 (overleaf left) Juan Patricio Morlete Ruiz, *The Sacred Heart of Jesus*, c. 1759, Mexico.

FIGURE 6.4 (overleaf right) Juan Patricio Morlete Ruiz, *The Immaculate Heart of Mary*, c. 1759, Mexico.

classical traditions and, most notably, Aristotelian constructs. It is also striking that the heart has been completely abstracted from the body. It has been afforded an identity and potency that is independent of the host body, and yet it is entirely representative of the host.

Despite the various embellishments, this style of depiction of the heart is clearly lifted from the world of the anatomists. One might question why. To our modern eye, this confluence of the medical and the sacred does not necessarily sit well. Some might even feel some discomfort in seeing the graphic 'scientific' image in this religious context.

Remarkably, we can trace the origin of this specific iconography to a single source, a hundred years earlier: the visions and subsequent letters of Marguerite-Marie Alacoque, a Visitandine French nun in the convent of Paray-le-Monial, close to Lyon.* In 1675, she told how, during the Eucharist, she witnessed a miraculous vision of Christ, who literally revealed his heart to her. In relating her experience, she was reviving a tradition of mystic somatization of the heart that went back to figures who we have considered previously: St Catherine of Siena, St Ignatius of Antioch, St Chiara of Montefalco and St Augustine. So, while the core ideas were not new, Alacoque's vision eventually attained much greater prominence than those of her predecessors and spawned the modern Cult of the Sacred Heart that would spread across the world. The florid representations of her vision were to play a crucial role.

In her vision, Christ had provided remarkably specific instructions to Mary, not least in relation to the subsequent depiction of his sacred heart and the dissemination of its image. Firstly, as recalled by Alacoque, Christ revealed his *actual* heart in the context of the Eucharist and asked that devotions should be directed specifically to the heart. Secondly, in

---

* Visitandines were a Roman Catholic order of nuns founded by St Francis de Sales and St Jane Frances de Chantal in 1610. Originally the order was intended to undertake charitable work, caring for the poor and the sick in their homes, but after only a few years, the founders accepted the more conventional requirements for closed orders and cloistered life.

her letters and in her autobiography, Alacoque recorded certain very specific features of Christ's heart, as he revealed it to her:

> I saw this divine Heart as on a throne of flames, more brilliant than the sun and transparent as crystal. It had Its adorable wound and was encircled with a crown of thorns, which signified the pricks our sins caused Him. It was surmounted by a cross which signified that, from the first moment of His Incarnation, that is, from the time this Sacred Heart was formed, the cross was planted in It...[7]

Crucial for the wider dispersal of Mary's vision was her conviction that Christ had instructed her to instigate a feast day to honour his heart – a cause that was taken up and propagated by the Jesuits. Mary's Jesuit confessor, Father Claude de la Colombière, was succeeded by Jean Croiset, who encouraged Alacoque to write an autobiography, which she did in 1685. Six years later, Croiset published his text *Sacré coeur de notre seigneur Jesus Christ* (*Sacred Heart of Our Lord Jesus Christ*) (1691), which incorporated Alacoque's biography, together with a description of what devotion to the Sacred Heart entailed. In 1704, it was placed on the church's Index of Prohibited Texts, but it attained widespread distribution nonetheless, with the publication in Rome of Joseph de Gallifet's (1663–1749) Latin text *De Cultu Sacrosancti Cordis Dei Ac Domini Nostri Jesu Christi* (*On the Worship of the Sacred Heart of God and Our Lord Jesus Christ*) (1726). This was the first devotional text about the Sacred Heart to receive widespread attention.[8] Alacoque herself had penned a primitive sketch of the heart as she had seen it, but Gallifet's book contains what was to be the most influential depiction.

Like Alacoque, Gallifet was motivated to propagate the image of the Sacred Heart. Born near Aix-en-Provence, he was admitted at the age of fifteen to the Jesuit Order, where he encountered Colombière. While on a mission of charity during his third year of probation at Lyon, Gallifet became gravely ill. Fearing for his life, his fellow monks vowed in his name that, if he were spared, Father de Gallifet would spend his life in devotion to the Sacred Heart. From the time he dedicated himself to that cause, he is said to have recovered, and afterwards he gratefully ratified the vow. Gallifet undertook three successive rectorships and, in 1723, he was promoted to be the assistant for France, an office that took him to Rome. His book *De Cultu* appeared there in 1726 and made a persuasive case for the establishment of a feast for the Sacred Heart, thereby fulfilling its author's pledge.

Gallifet's arguments were rejected by the church for decades, but the feast was eventually granted in 1765. While Gallifet did not live to see that goal realized, his influence was still far-reaching and led to the establishment of 700 confraternities of the Sacred Heart.[9]

But our main interest is in the pair of illustrations that form the frontispiece of *De Cultu* (Figure 6.5 and 6.6). On the left is the Sacred Heart of Jesus and on the right is the Immaculate Heart of the Virgin Mary. The left-hand image is a precise and literal rendering of Alacoque's vision. Beside the radiant, wounded hearts, the paraphernalia of crucifixion and the adoring attendant cherubim, the standout feature is undoubtedly the graphic anatomical nature of the hearts themselves. As Lauren Kilroy-Ewbank notes in her essay 'Holy Organ or Unholy Idol? Forming a History of the Sacred Heart in New Spain' (my emphasis below):

> Charles Natoire's influential engraving, included within Gallifet's 1726 Latin text, followed Alacoquian iconography closely, but departed radically in one manner from all earlier depictions of the Sacred Heart: *he shows Christ's heart as an anatomically correct organ.* Natoire employed subtle gradations in shading to emphasize the meaty carnality of the heart, convincing viewers that this is an actual eviscerated human organ.
> Gallifet's widely distributed book thus contained one of the earliest representations of Christ's heart to combine the mystical with anatomical naturalism—a combination visualized consistently

in Novohispanic paintings. Outside of the print genre, most European depictions of the Sacred Heart do not adopt this intense naturalism, perhaps a direct result of the long-standing emblematic tradition of showing the heart as a scallop-shaped organ, conceptualized visually in works like those of the Wierix brothers' *Cor Iesu amanti sacrum* (*c.* 1600) or Benedict van Haeften's *Schola cordis* (1629).[10]

Indeed, comparison of the Natoire hearts with other near contemporaneous depictions is revealing. Anton Wierix produced a series on the 'Devout Heart'. From the 1620s through to the nineteenth century, various artists reproduced the Wierix series, whether by directly copying the designs or drawing upon the format. These *Cor Iesu* images were printed either as part of a book or as loose sheets to be used as devotional imagery. The powdery, sentimental, perhaps even kitsch images of the Wierix series show the dove, cloud and cherubim in cartoon-like form (Figures 6.7 and 6.8). Note, though, that the heart itself is still in the motif form. It is plump, soft, smooth and utterly banal when viewed beside the disturbing Natoire and Morlete Ruiz versions.

In fact, Natoire had styled his highly anatomized version with good reason. One might have imagined that, armed with a knowledge of the new anatomy, the illustrator had taken it upon himself to depict the heart with anatomical precision, perhaps to demonstrate his newfound knowledge or technical skill. But this seems not to have been the explanation. In a letter to Croiset, Alacoque stated that Christ

FIGURE 6.5 (overleaf left) Charles Natoire, Sacred Heart of Jesus from the frontispiece for *De cultu sacrosancti cordis Dei ac Domini Nostri Jesu Christi*, 1726, Rome.

FIGURE 6.6 (overleaf right) Charles Natoire, Immaculate Heart of Mary from the frontispiece for *De cultu sacrosancti cordis Dei ac Domini Nostri Jesu Christi*, 1726, Rome.

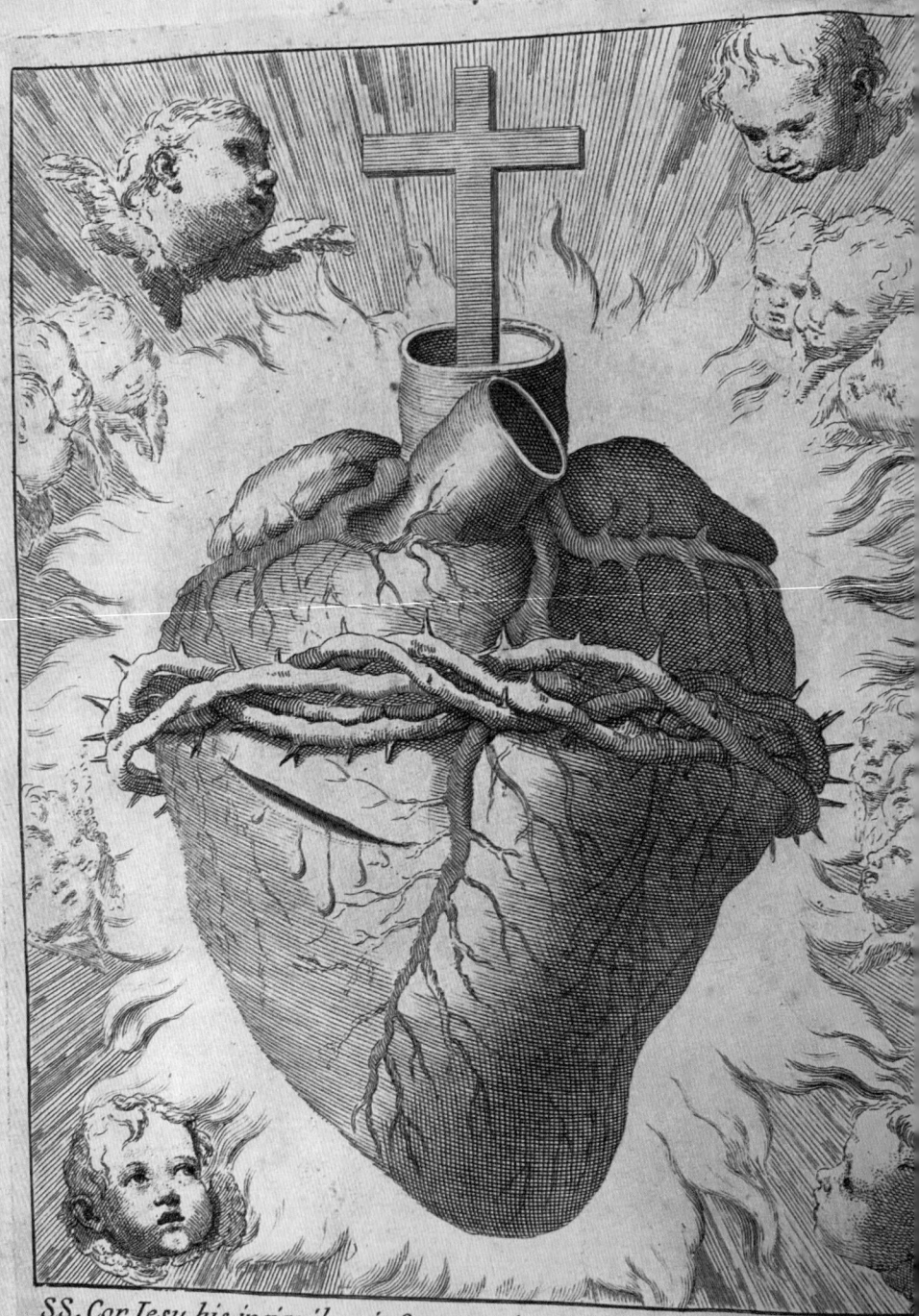

SS. Cor Iesu his insignibus instructum in visione Cœlesti Ven: Mat: Ma
Alacoque Ord. Visit. B. V. M. monstratum est.
Placuit autem SS. Cor in sua naturali forma ac magnitudine qualis in
mano corpore esse solet sculpi facere

Carolus Natoyr del.

Petrus Massini S

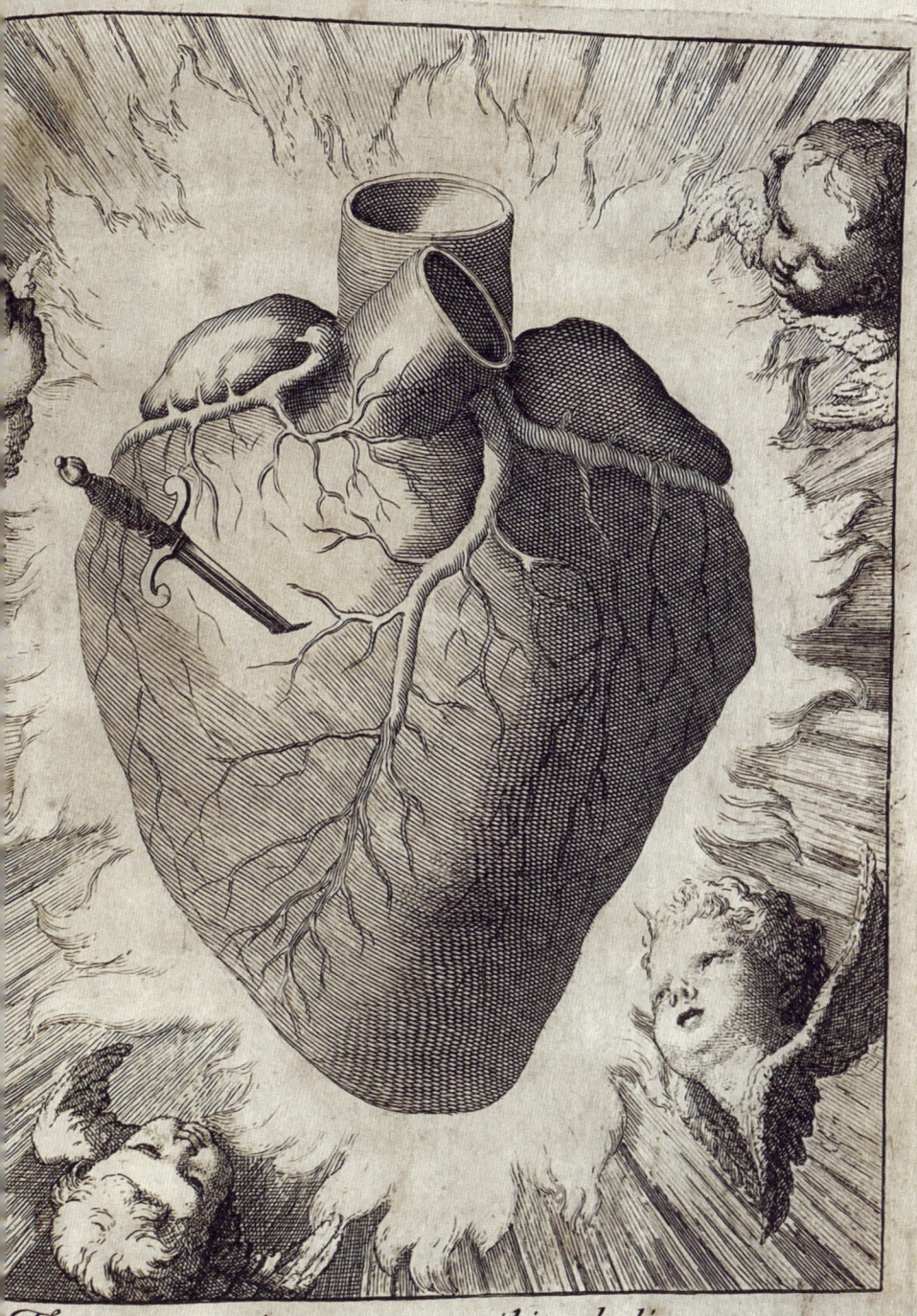

Tuam ipsius animam pertransibit gladius — Luc. 2 —

MARIÆ immaculatum amore Iesu quo ardes inflamma
Cor nostrum

Natoyr del.   Petrus Maßini sculp.

Eia IESV tibi notum.   An non cernis? tota patet
Cor, si lubet, lustra totum.   Ara cordis, nil te latet:
Pia tuo sanguine.   Foue tuo lumine.

Anton. Wierx fecit et excud.

Sat est, IESV, vulnerasti,
Sat est, totum penetrasti
Sagittis ardentibus.

Procul, procul hinc libido:
Nam cælestis hic Cupido
Vincet ignes ignibus.

Anton. Wierx fecit et exc.

had expressly commanded that his heart be displayed, as a public demonstration of the fulfilment of his requirement to promote the Devotion. Alacoque also described the explicitly visceral nature of the heart image. The result was a transition in heart iconography, a shift from motif to graphic representation:

> It must be honored under the symbol of this Heart of flesh, whose image He wished to be publicly exposed. He wanted me to carry it on my person, over my heart, that He might imprint His love there, fill my heart with all the gifts with which His own is filled, and destroy all inordinate affection. Wherever this sacred image would be exposed for veneration He would pour forth His graces and blessings.

The legend inscribed under images of the heart suggests that Natoire was conscious of the stylistic departure, since he seems to have felt the need to explain, perhaps even justify, his adoption of this radical new format:

> SS Cor Iesu his insignibus instructum in visione Coelesti Ven:
> Mat. Marg. Alacoque Ord. Visit. B.V.M. monstratum est
> Placuit autem SS Cor in sua naturali forma ac magnitudine
> qualis in humano corpore esse solet sculpi facere
> Carolus Natoyr del.
> Petrus Massimi sculp.

(The most Sacred Heart of Jesus, furnished with these symbols, was shown in a heavenly vision to the Venerable Mother

FIGURE 6.7 (previous pages left) Anton Wierix, *The Christ Child cleans the believer's heart with a hand-held mop, assisted and venerated by angels*, c. 1600, Antwerp.

FIGURE 6.8 (previous pages right) Anton Wierix, *The Christ Child shooting arrows into the believer's heart to conquer the fires of lust*, c. 1600, Antwerp.

174

Margaret Alacoque, of the Order of the Visitation of the
Blessed Virgin Mary
It was decided however to have the sacred heart depicted in its
natural shape and in the same sort of size that it usually has in
the human body
Carolus Natoyr drew.
Petrus Massimi engraved.*)[11]

Clearly, these Natoire representations of Marie Alacoque's vision
were the direct progenitors of the Morlete Ruiz paintings. Natoire's
images of the Sacred Heart would have made their way to the Spanish
colony of Mexico in the form of devotional texts and looseleaf prints.
Gallifet even made reference to the intentional propagation of images
of the Sacred Heart for the purpose of spreading the Devotion:

> We already have the consolation of seeing this picture in thou-
> sands of places, everywhere it is the delight of pure souls, and we
> have a firm confidence that it will daily become more common.
> This ought to be one of the chief interests of those who are devout
> to the Heart of Jesus: they should zealously endeavour to obtain
> this glory for the Sacred Heart.

---

* The inscription on the second page reads:

Tuam ipsius animam pertransibit gladius – Luc. 2
Cor Mariae immaculatum amore Iesu quo ardes inflamma.
(A sword will pierce your own soul – Luke 2
Immaculate Heart of Mary, inflame our heart with the love of Jesus with which you
burn.)

The relevant biblical verse is:
(Yea, a sword shall pierce through thy own soul also,) that the thoughts of many hearts
may be revealed.

Luke 2:35, King James version

They should have the picture exposed for veneration with the greatest possible magnificence in churches, in houses, in private oratories. They should carry it about with them as a precious token of their love for Jesus Christ.[12]

Indeed, Morlete Ruiz was not especially known for sacred art, but he was known for recreations of European works in oil, including paintings of European landscapes derived from a series of engravings. Prints popularizing the Sacred Heart had quickly been disseminated in France, Italy and Spain, drawing heavily on the iconography derived from Alacoque. Gallifet's book itself was translated into vernacular languages, notably French (1733) and Spanish (1734) by the Jesuit Bernardo de Hoyos. These translations ensured the widespread dissemination of *De Cultu* across the Catholic world, as did the Jesuits' endorsement of the emerging cult. In fact, through its affinity with visions and mystical Catholicism, the Society of Jesus became intimately connected to the Cult of the Sacred Heart.

Through a monopoly granted to the Plantin Press in Antwerp by Philip II in about 1570, huge quantities of illustrated religious books and prints were sent to Spain for export to her colonies. The trade continued for almost 250 years, exporting stylistic influences to Mexico from Europe, including those of Maerten de Vos, the Wierix brothers and, later, the baroque works of Peter Paul Rubens.

The persistence of the naturalistic versions of the heart is particularly interesting, not least because, to our modern eye, they are so jarring. Here is an anatomical image, a graphic image, thrust into a context in which we might expect a gentler, sublime, *holy* image. But that is the point. By making the heart of Christ fleshy (recall the fleshy tablets of the Vulgate and the bleeding heart of St Augustine), the wounded, violated heart shown in Morlete Ruiz's painting *is* the heart of Christ. Were we to doubt this, he has made the message unequivocal by showing a punched-out disc in Christ's heart. In the Catholic Eucharist (since the thirteenth-century doctrine of transubstantiation), the bread does not merely represent the body of Christ, it *is* the body of Christ.

Although the images of the heart were widespread in the Novohispanic world, in particular through dissemination by the Jesuits, it is worth emphasizing that even at that time, this was cult activity and not without controversy. Even recognizing the instinctive appeal and the potency of the symbolism attached to the Sacred Heart of Christ, there was no scriptural basis for the devotion. Indeed, some considered it heresy and the veneration of the heart to be a form of idolatry. To bolster its validity and to undermine potential objections, proponents invoked a tradition of Sacred Heart-related theology. The original Natoire rendering places the floating heart centrally and it utterly dominates the plate. The cherubim are both anonymous and secondary. However, in the Morlete Ruiz versions, these elements are placed alongside identifiable luminaries of the Catholic past, which are configured around the heart, implying a lineage that lends credibility to this theology and marking an attempt to legitimize it by weaving it into the Catholic mainstream. Hence, in Morlete Ruiz's image of the Sacred Heart of Christ, and moving clockwise from God the Father at the apex, we find Philip Neri (not Society of Jesuits [SJ]), Francis Xavier (SJ), Stanislaus Kostka (SJ), St Francis de Sales, St Francis and St Bernard of Clairvaux (beneath John, Mary and Mary Magdalen), St Augustine, Luis Gonzaga (SJ), St Ignacio of Loyola (SJ) and Francis de Paola.*

Christ's heart and accoutrements embodied the religiosity of the Catholic Church, with its arcane mystical practices. The Jesuits, always active promoters of the devotion of the Sacred Heart, adhered to such practices to access and experience the divine. Indeed, the Sacred Heart came to symbolize the very identity of the Jesuit Order. In his essay on Loyola, the French literary theorist Roland Barthes emphasizes the necessary materiality and physicality: 'The body in Ignatius is never

---

* I am most grateful to Dr Eric Southworth for these attributions.

conceptual; it is always this body.' In relation to depiction, Barthes states that: 'Semiologically, the image always sweeps beyond the signified toward the pure materiality of the referent.'[13] In other words, the heart is not merely denoting an idea, it contains meaning in its own physical identity, which rings true not only in this specific context but in our broader considerations of how depictions of the heart have been deployed in both secular and sacred contexts.

Despite its evident appeal and rapid expansion, there were also many critics of the cult, including religious reformers such as the Jansenists. Notable among their number was the mathematician Blaise Pascal (1623–62). He associated with this reformist movement named after Cornelius Jansen (1585–1638). Following Augustine, the Jansenist outlook was more austere, theologically grounded and scientifically informed than the Jesuits. They were opposed to what they regarded as the unfounded mysticism around the Cult of the Sacred Heart.[14] Following (perhaps despite) his own intense religious experience of the presence of God, Pascal seems to have taken a more intellectually humble and open-minded view, simply accepting that: 'The heart has its reasons that reason does not know.'[15]

Some Enlightenment thinkers invoked the emergence of the new physiology expounded by Harvey and his followers in order to downplay the spiritual significance of the heart. Detractors pointed to the work of Harvey (*De Motu Cordis*) and others as rationalizing the role of the heart in the newly described system of circulation.[16, 17]

By the mid-1760s, Charles III of Spain – concerned by the growing influence of the Jesuit Order, its political alignment with the papacy and what he saw as its undermining of the authority of the Spanish monarchy – expelled the Jesuits from Spain and the Spanish colonies and called for the destruction of images of the Sacred Heart. But the Sacred Heart survived and flourished nonetheless, and by the mid-nineteenth century the cult had spread across the globe, from France, Italy and Spain to Peru, Colombia and the Philippines. Some of the most complex and iconographically rich depictions to survive come from Novohispania (Mexico and Central America north of Panama, the Spanish West Indies and the Philippines), but more modern cathedrals and basilicas dedicated to the Sacred Heart can be found as far afield as Yokohama (Japan), Pondicherry (India), Brazzaville (Democratic Republic of Congo) and Harare (Zimbabwe).

The elaborate and formal paintings of Morlete Ruiz were undertaken in the late eighteenth century. They were costly works, executed in oil on copper, commissioned by the wealthy and highly polished in their style. They were sophisticated, iconographically correct and politically calculated, but they were essentially embellished copies of European engravings. By the end of the eighteenth century a style of vernacular art had emerged in Mexico that, while not entirely insulated from formal artistic traditions, was more expressive and free-flowing.[18] Paintings in this style were made on tin plate, a versatile, resilient and democratizing medium on which to paint. The term 'retablo' refers to one form of such paintings, reflecting their original devotional purpose. They were positioned *retro tabulum* ('behind the altar'), where they superseded the formal painted or carved gilded screens that were traditionally placed there, which typically depicted scenes from the lives of the saints.

The Academy of San Carlos in Mexico City, which would have exerted an influence on more formal styles, had little reach into the folk art of the retablo painters, who operated unburdened by the weight of art history and unbound by the constraints of the academic art establishment. Nonetheless, their iconography was predetermined by the church, and by its semi-rigid religious canons, imported through illustrated religious books and prints.

Such paintings could serve as objects of veneration on an altar in the home. They often portrayed saints, the manner of whose depiction was copied from pictures or statues in churches, from books that contained engravings or books of daily devotion. Retablos, however, have a character of their own: they were typically conceived and created by devout, mostly untrained craftsmen who, although striving for faithful replication of a 'correct' sacred image, interpreted their subject through the prism of their own circumstances and cultural beliefs, and within the parameters of their technical ability and resources (Figure 6.9).

It would be wrong to give the impression that the Sacred Heart was a dominant image in retablos, but it was undoubtedly prominent, not least in the form of Our Lady of Sorrows (*Mater Dolorosa*). Although presented with minor variations, the common theme in that scene was of Mary, mother of Jesus, as the mourning mother. She is sometimes crowned with thorns over her veil, but she usually appears in a grieving, sometimes tearful attitude, with her hands clasped. There is always a dagger or similar bladed weapon in her breast, and sometimes seven daggers, reflecting the punctured Immaculate Heart discussed earlier in this chapter and the allusion to the account of how, when presenting Christ to the temple, Mary was told that a sword would pierce her soul (the prophecy of Simeon, in Luke 2:35). This portrayal has been further elaborated to include the 'seven sorrows of Mary', which include scriptural and apocryphal causes for her sorrow (not only the crucifixion, but also the piercing of Jesus' side and the burial by Joseph of Arimathea). The dominant feature of the *Mater Dolorosa* is always the piercing of the breast or heart, specifically symbolizing the death of Christ.

Applying the term 'retablo' to the form of sacred art described above enables delineation of a separate but related tradition of 'ex voto' or votive paintings that depict dramatic events in the lives not of saints but of ordinary people, and which were created to give thanks to God, Mary or a saint for some perceived miraculous intervention that averted otherwise certain disaster. These were naïve pieces, undertaken by craftsmen rather than trained artists. Ex voto paintings might show a person prostrate with illness or injury, and family members gathered

FIGURE 6.9 Miguel Claflin, *Retablo* (Sacred Heart), 1939.

MAJEL G. CLAFLIN - TAOS - N. M.

around, praying for that person's recovery. Alternatively, they might depict an accident or violent episode that the victim was fortunate to survive, and for which deliverance, therefore, thanks were due. Occasionally, the artist might attempt to create a likeness of the principal characters, but more often their physical characteristics simply reflect the style adopted by the artist.

Often, the saint whose intervention had been called upon, and to whom the victim owed their salvation, would also appear in the painting. The narrative depicted in the image is usually accompanied by a few lines of explanation and an offer of thanks. Although such ex votos might be painted by the person who had experienced the miracle, they were more often taken on as commissions by individuals from the same locality. The client would relate their experience and the artist would interpret this in drawing and paint. The costume and personal detail would receive lavish attention, along with a careful narrative of the event. The backgrounds were often sparse, in a single field with muted hues of buttermilk, ochre and indigo. Any attempt at perspective was crude.

The completed ex voto would be hung on a church wall or placed near a particular image to commemorate the recovery of the donor, and it was neatly termed 'a receipted bill for spiritual or physical boons received'.[19] In a society where many of the poor could not read, these graphic depictions were accessible and powerful articulations of deeply felt gratitude. As Roberto Montenegro, himself a noted Mexican muralist, noted: 'The lack of technique, the discretion of the use of tone and inimitable charm of the fashion of the time, created a school which affords us a sensation of sincerity.'[20]

The period that saw the greatest production of ex voto paintings was concurrent with the period of retablo production, roughly from the end of the eighteenth century through the nineteenth and early twentieth centuries. Before the twentieth century, these votive paintings were always anonymous; the retablo belonged to the person represented and was a form of sacred communication. It was not a creation instigated by the artist or intended for his glorification. Moreover, the work was most likely undertaken on a fee-for-service basis by a painter working as a tradesman or artisan.

> The work of a native artist or anonymous painter mirrors his naive soul. Although he is contending with technical and compositional handicaps, the spirit expressed in his work and its story-telling qualities are unique.[21]

Allow me to describe a typical, yet highly specific, example. In the main scene a young woman lies across a tramline. There has been a collision and the woman is seriously injured. The *Mater Dolorosa* is suspended in the top left-hand corner of the image; her facial expression is laden with sorrow and her heart is encased in a crown of thorns and pierced with a dagger. At the bottom an inscription reads: 'Mr. and Mrs. K— give thanks to Our Lady of Sorrows for saving their daughter F— from the accident which took place in 1925 on the corner of Cuahutemozin and Calzada de Tlalpah.'

Produced on standard-size tin plates, these votive pieces were commonplace in Mexican society. Their dramatic and highly narrative composition and rectangular uniformity almost have the quality of movie stills. Taken together, the ex votos give a sense of the dramatic history of the communities in which they were created.

> From place to place and period to period, significantly, occupations, situations, official clothing, progress in caravan against a changeless endless background, vibrant of human trouble and of racial agonies throughout. Plagues, droughts, conflicts, are dated and described. The very emotion concurrent is charted, in kind and quality. In the quiet of miracles some years, the violence others [*sic*]; in the faith that makes them numberless. It is a moving record of a nation, a stethoscopic measure of its heart.[22]

7

The Modern Heart

[The] only example in the history of Art, of someone who rips open their breast and heart to speak of the biological truth of that what they feel inside… The only woman who has expressed in her works of art the feelings, the functions and the creative potential of women…

DIEGO RIVERA

1943

In the story so far, we have explored how common themes in heart depiction have stretched over apparently disparate historical and geographical landscapes. We have seen ideas of the heart as the seat of the soul and the source of emotion and witnessed tortured hearts in love and grief. Hearts have been inhabited by gods and inscribed with the records of deeds or with holy texts. They have been seized and squeezed and bled and exchanged. Despite the breadth and variety of connotations, the consistency of certain intuited properties associated with the heart is remarkable, and the common attributions seem to underpin the heart's prominence in images and the shared, often elaborate ways in which the heart has been depicted.

And yet, from the seventeenth century onwards, the heart was recognized as a muscular pump at the hub of a system of branching tubes. The circulation of the blood had been elucidated and its prosaic purpose laid bare. For all of its pre-eminence as the first moving structure in the embryo, the elegance of its design, the precise architecture of its spiralling muscle fibres and the valves that choreograph the internal dynamics of competing blood flows, the heart was, in the end, a pump.

Once the very literal function of the heart had been determined, one might imagine that the earlier intuited attributes would be, in some way, diminished; the mystery lost. On the other hand, if we engage with the felt experience of our own hearts sufficiently deeply, perhaps knowledge of mere function hardly impinges on the sense of a deep vital presence that we simply cannot avoid. With that in mind, we might ask whether modern artists have used heart iconography and, if they have, in what ways modern usage has emulated (or departed from) the traditional versions.

Before the nineteenth century, artistic endeavour was biased by the commissions of wealthy private patrons and by institutions, most notably the church and other religious bodies; there are many examples of this among the images we have seen already. The subject matter was commonly sacred or a depiction of some ancient, widely recognized mythology, god, creature or event. Their style was judged against historic practice or contemporary aesthetic expectation (often both), and conformity with these norms was expected. By the end of the nineteenth century and the beginning of the twentieth, these bonds had been broken. Modernism brought greater emphasis on exploration, innovation and expression of the self in the visual arts. In keeping with the mood of an age that also saw the publication of Sigmund Freud's *The Interpretation of Dreams* (1899), some artists began exploring alternative repertoires of expression that incorporated more personal symbolism and iconography, often reflective of their subjective experiences.

In this new context, we might ask whether knowledge of the function of the heart lessened its relevance as a visual icon, or whether, in the context of the advent of a new freedom of expression, common themes emerged or persisted in relation to the heart. In other words – did the heart persist as an icon, and if so, did it take on new meanings, or were the old meanings and symbolism perhaps repurposed outside of the historic contexts? If there is merit in the idea that a common conscious experience of the heart underlies the shared set of intuited attributes, then we might expect that even modern artists who were free to express themselves subjectively would use, or perhaps reinvent, the established repertoire of heart iconography; that the attributes accorded by previous generations of philosophers, religious thinkers, writers, artists and poets would resurface in a new form, or be influenced by either conscious or subconscious sources.

To examine these possibilities, I am going to look at the work of a number of modern artists who used heart motifs prominently and sometimes centrally: Frida Kahlo, Edvard Munch, Pablo Picasso, Andy Warhol and Alexander Calder.

No artist has used the heart so prominently, or with such potent effect, as Frida Kahlo. Like Marguerite-Marie Alacoque, her vision of the heart was deeply personal and graphically anatomical. Kahlo used heart imagery repeatedly over several years and drew from an exceptionally complex amalgamation of sources that included pre-Columbian ceremonial practices, the sacred and vernacular Catholic art that suffused her environment in Mexico, alloyed with the cold literal illustration of medical textbooks. Without doubt, Frida Kahlo extended the register of her emotional expression using heart iconography.

> I paint my own reality… I paint because I need to and I paint whatever passes through my head without any consideration.[1]

Her 'own reality' was a relatively short life of extremes, lived in turbulent times in the company of powerful personalities. What passed through her head and into her art was a train of dramatic images, often in the form of self-portraits that depicted her weeping, bleeding, eviscerated and divided self. She transposed emotion, notably personal anguish, into art with extraordinary candour and a palpable intensity, coloured by the rich cultural environment in which she lived.

Kahlo was born in Coyocan, Mexico City in 1907 (three years before the Mexican Revolution), the daughter of a German father of Hungarian-Jewish immigrant descent, Guillermo Kahlo, and a native Mexican mother, Matilde Calderon. In early life she was raised as a Catholic and was undoubtedly exposed to the sacred paintings in the churches of Mexico City.

In 1925, aged eighteen, she suffered severe, life-threatening injuries in a traffic accident, which were to have a profound, perhaps defining influence on her life and works. Kahlo was travelling on a bus that collided with a trolley car. Hayden Herrera's biography quotes Kahlo's own recollection of the event:

> It is a lie that one is aware of the crash, a lie that one cries. In me there were no tears. The crash bounced us forward and a

handrail pierced me the way a sword pierces a bull. A man saw me having a tremendous haemorrhage. He carried me and put me on a billiard table until the Red Cross came for me.[2]

The account of her boyfriend Alejandro Gómez Arias describes the wooden bus bending under the slow sustained force of the trolley until its structure gave way abruptly, fragmenting the carriage and propelling the debris that would pierce Frida's body.

I remained in the train. Not Frida. But among the iron rods of the train, the handrail broke and went through Frida from one side to the other at the level of the pelvis.
[…] a man said 'we have to take it out'. He put his knee on Frida's body. When he pulled it out Frida screamed so loud that when the ambulance from the Red Cross arrived her screaming was louder than the siren.[3]

Her injuries were profound. She suffered multiple spinal and pelvic fractures along with broken ribs and a crush dislocation of her right foot. The iron rod itself had pierced the left side of her pelvis, emerging from her vagina.[4] Nursed under poor conditions, it is astounding that she survived this horror at all.

Before the accident, Kahlo had intended to study medicine, but during her convalescence she started to paint. Her mother installed a mirror above her bed, from which she began to paint the self-portraits that, above all, would characterize her work. Though she was well enough to leave the hospital and visit Mexico City within three months of the accident, its physical consequences persisted throughout her life; she suffered long-term pain and required numerous corrective surgeries, splints, braces and prostheses. Her plans to study medicine were wrecked. Her internal injuries are likely to have contributed to her multiple miscarriages.

The psychological impact of her suffering is inestimable. While I am anxious not to reduce Kahlo's work to an analysis of her emotional turmoil, it would be misleading to suggest that she did not intentionally draw on her own experiences. The art historian Paul Westheim has written: 'Frida belongs to no school, no current, no movement; she lives on the margin. She is an artist who came into her own by giving form to her visions and fantasies.'[5]

Kahlo was neither attached to nor constrained by any particular artistic movement, and it is fascinating that she drew on the iconography of the heart in some of her most important work. She created and adapted powerful images of the heart (and blood vessels) that were derived from a range of seemingly disparate sources. In doing this, Kahlo was able to blend a potent and original emotional palette, which she used to expose her most intimate self.

While still at school, she had renounced her Catholic faith and embraced communism. She joined the Mexican Communist Party in 1927, through which she met Diego Rivera, who was the most prominent artist in Mexico and who already enjoyed international recognition. Within months they were married, but not without controversy; she was twenty-three, he was forty-two and divorced. By 1931, they had moved to New York and from there to Detroit, Rivera having been commissioned to paint a history of Detroit industry on the walls of the Detroit Institute of Arts. It was in that city, when she was three and a half months pregnant, that she suffered a severe haemorrhage, associated with miscarriage, for which she was admitted to hospital for emergency treatment.[6] Her 1932 work *Henry Ford Hospital* is painted on sheet metal and recreates the Mexican ex voto style. She uses naïve forms on a largely unpopulated backdrop to convey emotional narratives in relation to the trauma of her miscarriage.

A naked Kahlo lies on a bed, with a large pool of fresh blood and clots around her hips. She stares vacantly, obliquely across the canvas with an oversized tear on her left cheek. In her hand, she is clutching a leash of diverging red strands that are attached to six objects, each of which is suspended in ambiguous space around her. The strands are red and thin. Are they blood vessels? The one extending from her hand to the umbilicus of the dead foetus would suggest so, but on closer examination, each is tied like a ribbon or streamer to the object at its distal end. Each object represents a part of her most immediate and painful reality. There is an anatomical model, seemingly of the female pelvis and gravid uterus; next to that is the foetus in typical pose, with flexed limbs and disproportionately sized head. A bulbous snail looks pointedly away from the scene towards Detroit in the distance; apparently Kahlo said that it reflected the painful slowness of the miscarriage, but one wonders if it also contains something of the absent and distracted Rivera. There is a brightly coloured but wilting orchid in the central foreground, representing the empty uterus, with Kahlo's own pelvic bones on the right. A hinged industrial object of some kind, perhaps a pipe-cutting machine, is placed at the bottom left, invoking the brutality of medically severed vessels. This was her first work on sheet metal and, in keeping with the ex voto style, Kahlo records the date and place of the event, but she inscribes them on the frame of the hospital bed rather than on a scroll at the base of the image.

It is a desolate and harrowing scene. From the languid indifference of the grotesque snail to the harshness of the pipe cutter and the fragile beauty of the wilted bloom, Kahlo records an inventory of excruciating pain. The distant industrial buildings on the horizon refer to the city of Detroit and the remoteness and absence of Rivera, who continued with his work there. For Kahlo, this must have been a time of fear and solitude in the harsh, mechanical environment of a hospital, in a country that was not her own.

The combinations of objects, floating on an otherwise sparse background, bring to mind the painted tin plates in Mexican vernacular art. In using the primitive style and the graphic anatomical elements, she is grounding the scene in something fundamental and invoking shared human experiences of loss, grief and isolation. In common with some of her other paintings of that time, she has started to use images of blood and blood loss, but she has not yet used the heart.

The bizarre, the grotesque and the deeply expressive all push the *appearance* of Kahlo's work towards that of the surrealists, who aspired to a form of expression that was unclouded by the control imposed by reason and who adopted Kahlo as one of their own. Kahlo, however, resisted any such categorization of her work: 'They thought I was a Surrealist, but I wasn't. I have always painted my reality.'[7]

Though superficially bizarre and dreamlike, in the end what Kahlo depicted was not the contrived psychological exploration of the surrealists, but a *reality* articulated in paintings that were laden with a thoughtfully constructed personal symbolism. Perhaps we find her self-analysis so painfully insightful, and so open, that we can only approach it by imagining that it came from a dream. Though she distanced herself from the surrealists in relating the origins of her work, she certainly knew the circle from the inside; André Breton was an admirer and had stayed with her and Rivera in Mexico. Indeed, enabled by Breton, in 1938 Kahlo exhibited around twenty-five paintings in a solo exhibition at the Julien Levy Gallery on East 57th Street, New York City, the same gallery that had shown Max Ernst, Man Ray and René Magritte. In so doing, she was aligning herself, at least superficially, with the surrealists, although in the fullness of time that movement attached itself more to her than she to it. At the time, *Vogue* magazine presented her work as 'a sort of naïve Surrealism, which she invented for herself... free from the Freudian symbols and philosophy that obsess the official Surrealist painters'.[8]

In the following years she moved between New York and Mexico during a period in which Rivera was conducting numerous extramarital affairs, including with Frida's younger sister. In 1939, after Frida returned from shows of her work in New York and Paris, they were divorced.

FIGURE 7.1 Frida Kahlo, *The Two Fridas*, 1939, Instituto Nacional de Bellas Artes Mexico, Mexico.

Paul Westheim's 1952 double-page article in *Novedades* (*New Arrivals*) has a large, grainy photograph of Kahlo at its centre. She sits in a high-ceilinged studio, at her easel, looking out, facing the camera. The orientation means that we cannot see the work in progress, but behind her, hung high and dominating the space, is her well-known painting *The Two Fridas* (1939) (Figure 7.1). Here, she deals with her marital crisis and separation from Diego. The painting shows two versions of herself. The Mexican Frida, in the traditional *tehuana* costume, is seated on the right, while the other Frida is wearing an elaborate, Victorian-style, colonial white wedding dress. The facial expressions are fixed mirror images staring blankly from the canvas, but it is the way that the hearts and blood vessels have been depicted that make this an extraordinary image. In the traditional Frida, the heart approximates to the anatomical form that she may have seen in the medical textbooks and in the Novohispanic sacred art that would have been commonplace in her environment. It is closed, static and bloodless; the artery held by Frida's left hand (image right) is dry. It seems to be attached to a small button of pallid tissue that is said to be a portrait of the young Rivera.[9] The two Fridas' hands touch awkwardly and a wiry blood vessel extends between the two figures. The direction of implied blood flow is necessarily ambiguous. The heart of the Frida in the wedding gown has been laid open. Hayden Herrera tells how Kahlo related the traditional version of herself to the woman Rivera once loved and the wedding dress Frida to the woman he no longer loves, and that the painting is a 'straightforward symbol of pain in love'.

The anatomized heart and vessels, the bleeding, the surgical forceps and the emotional sterility of the characters all speak of another highly medicalized image. Supported by her own comments, the conventional interpretation is that the painting reflects Kahlo's anguish and desolation following the break with Rivera. Hayden Herrera records Kahlo's partial interpretation:

> The fact that I painted myself twice, is nothing but the representation of my loneliness. What I mean to say is, I resorted to myself; I sought my own help. This is the reason why the two figures are holding hands. There is little more I am able to explain because all of the truly lively motives for this painting are, without doubt, subconscious.[10]

Perhaps another lurking pain was too great to articulate in words, for, besides the loss of Diego, a further excruciating absence is being portrayed here. A second branch of the artery that extends from the heart does not extend to the traditional Frida, but passes behind her right shoulder and is cut (or perhaps surgically clamped), resting on her lap. It is truncated close to the pelvis, with fresh blood dripping onto the whiteness of her skirts and blending with the red blossoms that decorate the hems. Without undermining the conventional view, an additional interpretation seems likely. The position of the bleeding vessel and the clamp evoke the *absence* of the foetus at the end of the umbilical cord, although the anatomy is, of course, bizarre and contorted. To understand the origins of the image, we must go back to her time in hospital in Detroit. After the haemorrhage that led to the miscarriage (and we should recall also her original intention to study medicine) she asked to be given medical textbooks. It would surely have been natural for her to want to understand the nature of maternofoetal circulation; after all, its failure had threatened her life and cost the life of her unborn child. Apparently her doctors refused, but Diego sourced for her a copy of the *Principles and Practice of Obstetrics* by Joseph B. DeLee.[11] Clearly, she must have studied the book in detail. The figure entitled 'Diagram of fetal circulation' (Figure 7.2) is strikingly similar in style to what Kahlo produced in *The Two Fridas*. She has opened up the heart, and the net-like structures in both the textbook and the Kahlo painting represent the sinuous cords of the heart valve apparatus. It is connected to long, tortuous blood vessels that supply the placenta and, by extension, the foetus (which is not shown).

In *The Two Fridas* we see an intriguing fusion of the punctured or injured heart as an intuited way to represent grief in keeping with the traditional (Catholic) motif of female, and in particular maternal, suffering. It is embellished with a literal and modern allusion to medical physiology. The result is a powerfully evocative carnal picture of desolation. It vividly brings to mind Rivera's remark in an article

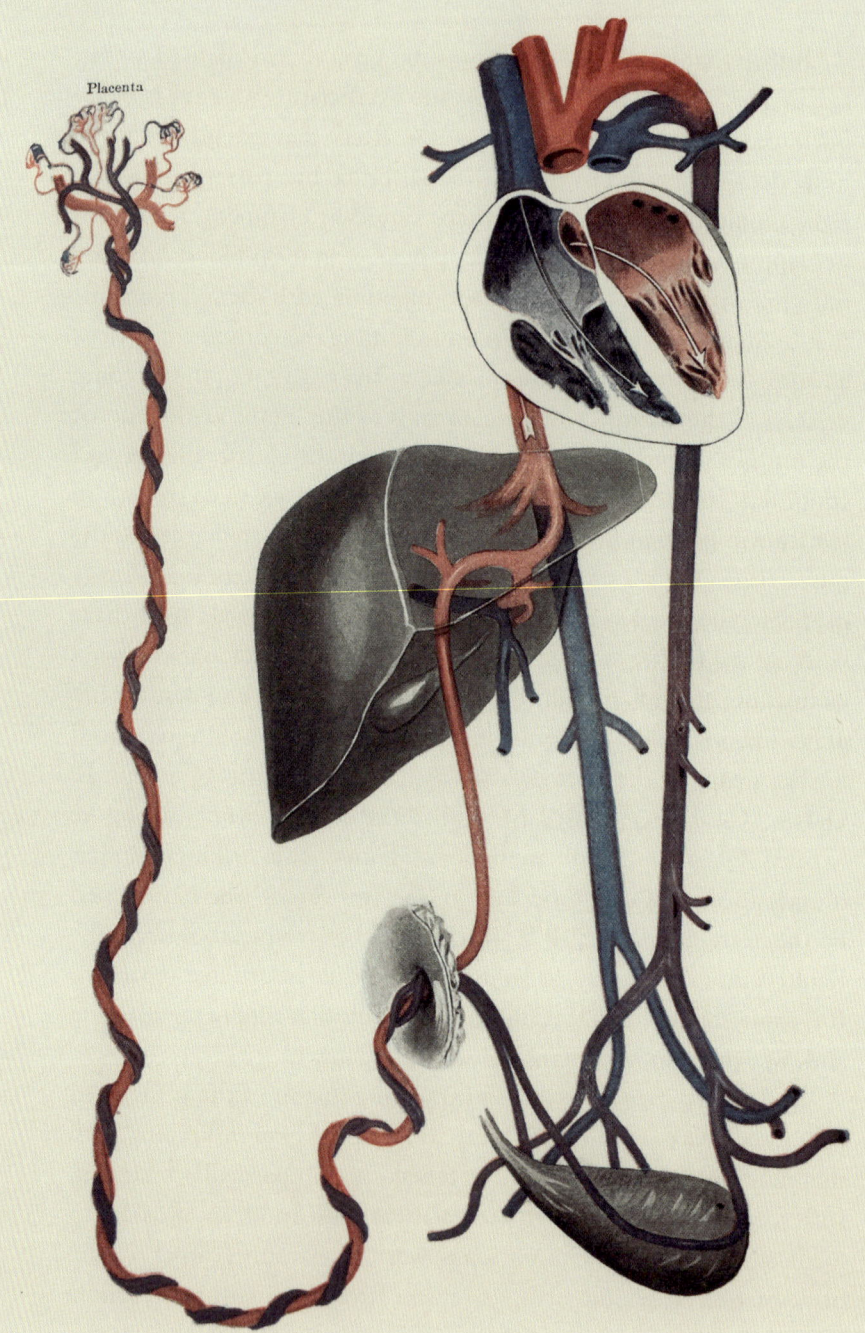

Placenta

published in the *Boletin del Seminario de la Cultura Mexicana* (*Bulletin of the Mexican Culture Seminar*):

> Frida is the only example in the history of Art, of someone who rips open their breast and heart to speak of the biological truth of that what they feel inside… The only woman who has expressed in her works of art the feelings, the functions and the creative potential of women…[12, 13]

Kahlo and Rivera had themselves accumulated a collection of many hundreds of retablos and ex voto pieces. The ex voto image described at the end of the last chapter relates in fact a reworked piece by Kahlo herself. Accidents of the sort that she suffered were so common that it seems she could source an existing ex voto piece, adjust the depiction of the injured woman to resemble herself more closely and rewrite the text to report the specifics of her accident and the gratitude of her parents. At the top left is Our Lady of Sorrows with the thorn-encircled heart derived from the original Alacoque visions.

The art historian Diego Sileo has described Kahlo's relationship with Mexican vernacular art as follows:

> It is precisely this correspondence with Mexican votive painting that alienates Frida's art from the Surrealism of Breton and his ilk. Frida's revisitation of indigenous folk art and her use of colonial era imagery allowed her to identify with her Mexican roots and reflect her desire to set herself in opposition to her contemporaries.[14]

FIGURE 7.2 Joseph B. DeLee, Fetal circulation from *Principles and Practice of Obstetrics*, 1913, Philadelphia.

Her 1939 work *Memory* (Figure 7.3), painted after her split from Diego Rivera, is modelled on this style. It depicts a river of blood seeping from the flaccid heart laying at her bandaged feet. Again, Kahlo is fusing the heart's properties of emotional intensity, and the cultural and religious motifs associated with it, with her personal experience of physical and emotional injury. The void in her chest is occupied by a metal rod that we imagine rocks rhythmically, painfully, spurred by the Cupid figures mounted astride each end. Clearly, the rod refers to her own excruciating physical pain, impaled on the metal handrail at the time of her accident, but it also alludes to the pierced hearts and chests that we have seen in the sacred and votive art of Mexico.

Her Catholic world would have been suffused with this sort of imagery and it is surely not necessary to try to pick out some specific trigger that might have prompted Kahlo to draw from what must been a large mental catalogue of heart images. But the *explanted* heart is a special case. It brings to mind the visions of the medieval nuns, notably Santa Chiara of Montefalco and, in particular, St Catherine of Siena, who exchanged her heart with Christ. In the treelined district of Coyocan, standing on the edge of a quiet garden square, less than a mile from Kahlo's home, is a church dedicated to St Catherine of Siena. The building is small; inside there are two rows of pews, with room for a dozen worshippers in each. The mustard-yellow, stucco-fronted church has a plaster-framed niche above the main entrance, from where St Catherine – dressed simply in white robes – gazes out; she holds a white lily in her left hand and her plump pink heart in her right. As in this tradition of sacred art, in *Memory* Kahlo chose to paint her anguished heart removed from her body. But it is not elevated or offered, or even clutched to her chest; it is abandoned, spewing, on the ground.

FIGURE 7.3 Frida Kahlo, *Memory*, 1937, Mexico.

Kahlo drew from pre-Columbian art in other works (for instance, in *My Nurse and I* and *The Love Embrace of the Universe*) and, given her propensity to fuse influences from disparate sources, it is plausible that similar influences were at play in *Memory*. Spanish texts from the sixteenth century, notably the Codex Magliabechiano, document the ceremonies of the Aztecs in which human hearts were excised (beating, from the living) and delivered to the gods.[15, 16] The Aztecs believed that the heart (*tona*) was both the seat of the individual and contained a fragment of the sun's heat (*istli*). Heart extraction was viewed as a means of liberating the *istli*.

Despite the rich possibilities for the provenance of Kahlo's hearts, ultimately the source (or sources) of her inspiration matter little. What is perhaps more interesting is that she chose to adopt the heart from the wide range of other icons that she might have used, and in so doing, she was drawing on traditions that have used images of hearts to convey the most profound human ideas and sentiments.

Were we in any doubt that Kahlo painted 'from the heart', a late work makes the point in literal fashion. In her *Self-portrait with the Portrait of Dr Farill*, she depicts herself beside the painting of her spinal surgeon. By this time her leg has been amputated; she sits in a wheelchair by the easel, on which is propped the completed portrait of Farill. On her lap she holds her palette, directed towards the viewer and moulded in the shape of the heart. It is an anatomized heart, with coronary arteries and veins, and it occupies the centre of the image. In her right hand she holds a clutch of paintbrushes dripping with blood.

It is hard to conceive how the heart icon could have been used more personally, more emphatically or distilled from more diverse sources. Kahlo grew a unique personal language from exceptional circumstances, incubated in her own particular environment. Although the range of artists who have employed images of the heart is large, Kahlo's depiction of the heart 'as organ' is exceptional, particularly as applied to the depiction of anguish. Her use of the heart emerged from a wish to reveal not just her story but her own response to it; a desire to depict emotion in painting.

Let us now turn to an artist with similarly stated intentions, but emerging from a very different tradition in Protestant northern Europe and with his own deeply troubled personal story. Edvard Munch (1863–1944) painted his best-known work, *The Scream*, when he was thirty years old and living in Kristiania (now Oslo). It is one of the most recognizable paintings of any era. By his own account, it depicts an actual event in which Munch recalls a walk that he took with friends. At some point, the group became separated and, while they were apart, Munch was overcome by awareness of an extraordinary skyscape, which left him in an anxious and agitated state.

He inscribed one version of the work:

I was walking along the road with two friends. The Sun was setting –
The Sky turned a bloody red
And I felt a whiff of Melancholy – I stood
Still, deathly tired – over the blue-black
Fjord and City hung Blood and Tongues of Fire
My Friends walked on – I remained behind
– shivering with Anxiety – I felt the great Scream in Nature.[17]

The person on the walk was Munch, but the person depicted on the bridge is intentionally amorphous, ambiguous. The painting depicts a state of terror. The wide-eyed, almost skeletal subject is fused through shared colour and the contour of line with the skyscape behind. The ordered linearity of the boardwalk and the figures in the background are disrupted by the curvilinear form in the foreground who, in motion and form, is subsumed by the intense fluxes of the natural world behind. In some sense, the personal experience or the specifics of the situation need not concern the viewer. Munch stated that the figure was recoiling from the 'scream of nature', but precisely what he meant by that is less clear and, perhaps, does not much matter. *The Scream* is so powerful because the anonymous form conjures a response from the viewer that is not so much one of empathy, as a powerful resonance at a frequency that evokes the viewer's own experiences and susceptibilities. Munch's triumph is to have conveyed some terrifying *internal* emotion that cannot be readily communicated in painting. Munch was in the process of rejecting the prevailing styles of realism and impressionism, the former with its emphasis on naturalistic depiction of the banal and the latter as an exploration of the effects of light on an *exterior* world, observed from a distance. He must have seen these as irredeemably superficial as he sought to look inwards, to explore human themes: love and jealousy, loneliness and anxiety, sickness and death. In so doing he was working towards the aspiration that he laid out in his *St Cloud Manifesto*, in which he declared that: 'No longer would interiors of people who knit and read be painted. There should be living people who breathe and feel, suffer and love.'

Like Kahlo, Munch was explicitly striving for a visual language with which he could articulate inner experience and its interpretation. In very different ways, they had both been exposed to life-changing adversity in early life. Munch's mother died of tuberculosis and eight years later, aged thirteen, he feared for his own life, having self-diagnosed his tuberculosis after coughing blood and clots. He recorded somewhat graphically:

> I was very frightened. I could feel the blood rolling inside my chest with each breath that I took. It felt as if the whole inside of my chest had come loose and was floating around, as if all the blood had broken free and wanted to rush out of my mouth.[18]

Mental illness was prominent in his family; his sister was committed to a psychiatric hospital, and he lived in fear of succumbing to mental illness himself.

By the time he reached his thirties, Munch was touring the salons and galleries of Berlin, Paris and Copenhagen. In 1889 he had been awarded a scholarship to study in Paris, where he was exposed to the work of Van Gogh, Toulouse-Lautrec and Seurat, among others, at the Salon des Indépendants. Their work influenced Munch significantly, as he edged towards his particular style of introspective symbolism. The art historian Allison Morehead makes a persuasive argument that he was further influenced by the intellectual environment of nineteenth-century France, including an emphasis on experimentation in the new science espoused by Claude Bernard, which had a profound influence on parts of the artistic avant-garde.[19] Applied to questions of human psychology, there was an emphasis on a study of the natural 'experiments' of altered states (such as delirium, psychiatric illness and sleep) to provide insights into psychological function. Morehead has written persuasively about Munch's ability to portray 'pathological' human states:

> The problem of many symbolists… as well as abstract expression-ists… was to create form that could be read as corresponding to an individual's, usually the artist's, interior emotive world, and to render that form objectively understandable, even truthful, without using what were understood to be or caricatured as traditional, academic, naturalistic modes of art making.
> More consistently than perhaps any other symbolist artist, Munch sought out pathological form for pathological content with a view to providing what he and his critical supporters imagined as truthful and universally legible representations of human experience.[20]

Munch developed a highly expressive iconography that incorporated the heart and blood prominently. He was preoccupied with ideas that revolved around suffering, death, anxiety, the pain of love, jealousy and neuroses of various forms. Much of Munch's work was self-consciously biographical – not in the superficial sense of a diary of events, but in the sense of a narrative of experience. In his 1896 painting *Separation*, Munch shows the desolation of that state. To the right, a featureless woman with flowing blonde hair trailing in the wind and a sweeping golden gown strides towards the canvas edge. In the foreground, a crumpled man gazes downward, his right hand clutching his chest, on the left side, over the position of the heart. The reddish hue on the edges of his fingers hints at the wounded, bleeding organ beneath. In front of him is an amorphous structure that some have likened to some crimson flora, or possibly flames. The form is ambiguous: others have suggested it may represent a mandrake, given that plant's association with love, death and mysticism, but what is depicted does not especially resemble the roots of the mandrake. It is more likely that the object, bilobed and with wavy edges, is a trembling, pained and discarded heart, quivering at the feet of the forlorn lover. This interpretation is plausible not only in the context of the wider composition, but because, as we shall see, Munch used the explanted heart device elsewhere.

In 1898 he had been commissioned to illustrate a volume of poems by his friend, the Swedish writer and painter August Strindberg. The poems were to be published in pamphlet form as a special issue of the German periodical *Quickborn*. The image in Figure 7.4 was first used as the cover of the pamphlet and Munch followed that publication with a series of related woodcuts. To our eye, these images are more than faintly melodramatic. In the *Quickborn* version, a troubled young man stares to the heavens, his neck fully extended, arched in a paroxysm of anguish with his flexed right arm raised to grasp his head. His muscular torso is tense and upright, but streaming through the fingers of his left hand, from an imagined wound or orifice, is a torrent of blood flowing onto the adjacent rock and towards the ground.

FIGURE 7.4 Edvard Munch, Cover for *Quickborn*, 1899, Pinneberg, Germany.

QUICKBORN

Blossom of Pain employs a similar device. Here, a crumpled man stares down at the earth. Again, his hand is placed over the chest as blood gushes through his fingers in an expanding stream onto the ground beneath. To the right of the man, limply draped over the edge of the rock on which he sits, is his own flaccid heart (Figure 7.5).

Elsewhere, in the woodcut *The Woman and the Heart*, Munch portrayed a young woman sitting naked on a lawn, among small scattered flowers (Figure 7.6). She sits with her back arched and her knees flexed, drawn towards her chest. She appears to be in a familiar pose, bathing. But instead of pouring water from a jug or squeezing a sponge, she is cradling an oversized heart in her outstretched arms.

FIGURE 7.5 (above) Edvard Munch, *Blossom of Pain*, 1898, Oslo.

FIGURE 7.6 (opposite) Edvard Munch, *The Woman and the Heart*, 1896, Oslo.

Her gaze is mostly obscured from us. The pastoral scene is calm, as she appears to drain a thin spout of blood, as if through a small hole in the anatomical apex of the heart. The blood drips onto her feet and forms a small pool.

In style, none of these images is at all like those of Kahlo, which, in a sense, makes their shared compositional elements all the more remarkable; each has used the flaccid, still heart at the foot of the emotionally wounded subject, with rivulets of blood ebbing out along the ground. It is hard to imagine any tangible link between the vastly remote worlds of these two artists and yet each alights on a common language of the explanted heart as an emblem of grief. A facile explanation, such as inspiration from a shared source, seems unlikely, since the way the two artists portray the hearts themselves is so different. Munch's is the almost unembellished heart motif, while Kahlo's is the carnal, anatomical heart of the Catholic tradition.

So how could it come to pass that Edvard Munch, the Norwegian son of a doctor, raised in a Protestant household in Kristiania at the end of the nineteenth century, and Frida Kahlo, a communist former Catholic steeped in Aztec art, Mexican ex voto painting and the sacred art of the Jesuits, converge on the use of the heart as icon?

The answer must surely lie in their common physical experiences of the heart and similar expressive purpose. Each was attempting to convey the 'unconveyable'; each was striving through painting to communicate some aspect of their own emotional state. Where Kahlo spoke of painting her painful 'reality', Munch wrote: 'My sufferings are indistinguishable from me, and their destruction would destroy my art.' He continues:

> I am making a study of the soul, as I can observe myself closely and use myself as an anatomical soul preparation. The main thing is to make an artwork and a soul study, so I have changed and exaggerated, and have used others for these studies.

[…] Just as Leonardo studies the recesses of the human body and dissected cadavers, I try from self-scrutiny to dissect what is universal in the soul.[21]

Similarly, while the composition and style of Kahlo's work bore some superficial resemblance to surrealist paintings, she was consciously articulating her own response to a personal history. Unlike the surrealists, she was not dredging silt from the murky channels of the unconscious mind – for example, through the exploration of dreams – but using painting to articulate her own *conscious* narratives. In so doing, she chose to employ a certain language that accorded heart iconography a prominent position. The style of representation of the heart that she followed, and the contexts in which she placed her depictions of that organ, reflected the breadth of her personal experience.

British artist Tracey Emin has also engaged with the heart as an icon with which to convey emotional intensity and anguish. In striving for depth of personal expression, she has been much influenced by Munch throughout her career. In her 2018 composition *Open Heart*, she seems to borrow from Munch directly. Here, a young woman with a featureless face gazing downwards kneels on a patch of rust-red grass. Her explanted heart hovers before her, loosely sketched in blood-red paint that drips down the image. The whole image is a composition of uniformly vertical dripping lines that drags the viewer's gaze and emotional response downwards.[22]

Among the dozens of prominent artists who have drawn on the heart icon, Kahlo and Munch stand out since they have not merely employed an established shorthand. Both of them have consciously cultivated images of the heart that draw on commonly intuited notions, while embellishing them in their own particular language and style. They do not use the brain, the kidneys, the liver, the pancreas or the spleen. Their choice of organ is the heart and only the heart. If they had used one of those other organs, imagine how little empathy we would have with what they were trying to convey. An image of a woman bathing in a meadow as blood seeps from a kidney cradled in her outstretched arms would be perplexing. We intuit something of the meaning intended by the use of the heart because we also speak its language. In their choice of the heart in the quest for a vocabulary of anguish, Kahlo and Munch are aligned with the punctured, crushed and pierced hearts of the Frau Minne images and the visions of Marie Alacoque. But we do not need to propose some intellectual lineage – I do not believe that the heart images in the work of Kahlo and Munch are an allusion or homage to earlier works. They are embraced as part of a common understanding. They are heartfelt.

Besides the expressive use of the tortured, anguished heart favoured by Kahlo and Munch, there are plenty of examples of hearts used to express love or passion more broadly. Pablo Picasso used multiple versions of the heart, particularly as shorthand to denote romantic love. When his lover Marie-Thérèse kept a record of their years together, he encapsulated her text record of the passing of each year of their relationship in a simple pencil love heart. More precisely, he chose the sacred-style heart surmounted by a plume of spirit and pierced by an arrow. Whether from his youth in Barcelona or his early years in Paris, living in Montmartre beneath the dominating Basilica of the Sacred Heart, Picasso must have had constant awareness of the heart and its presence in Catholic religious imagery. In his December 1931 painting *Femme au fauteuil rouge* (*Woman in a Red Armchair*), the heart motif replaces the head of the subject, Marie-Thérèse, and is also echoed in the form of the arms curving round the torso to meet at the apex. His 1945 work, *Femme dans un Fauteuil* (*Woman in a Chair*); afterwards known as *Femme au Buste en Coeur* (*Woman with a heart-shaped bust*), has the voluptuous heart not merely occupying, but comprising, the entire torso (Figure 7.7). The heart

FIGURE 7.7 Pablo Picasso, *Femme au buste en coeur* (*Woman with a heart-shaped bust*), 1945.

symbol is not particularly prominent in Picasso's enormous *oeuvre*, but its presence here illustrates the readiness with which the symbol can be used as a shorthand to convey romantic – and sometimes erotic – love.

The use of the heart motif for this purpose has become commonplace, almost trite. The wide dissemination of the heart as a symbol of love came about in the nineteenth century as a result of the popularity of valentine cards in the West. It is now so commonplace a symbol as to be almost uninteresting for discussion, and yet the very readiness with which the heart has been embraced for this purpose is significant. We simply don't attribute the emotions or physical feelings of love to any other organ.

We saw the first use of the 'love heart' in medieval depictions of chivalric love in Chapter 2. In Figure 2.12, the suitor's heart was removed and used to decorate – in various tortured forms – an image of Frau Minne, the unattainable object of his desire. In 1950s New York, where homosexual love was still necessarily covert – and in that sense unattainable – Andy Warhol employed a similar device. In his *Studies for a Boy Book*, he repeatedly adorned the lips, fingertips, bodies or genitals of young men with tiny, delicate heart motifs (Figure 7.8). In some sense, Warhol as lover was personified in the isolated heart, and the love-heart motif used to claim or badge the object of desire. In his essay accompanying Warhol's published collected drawings of this period, New York artist Michael Dayton Hermann observes:

> The intoxicating thrill of falling in love blinds us to the fact that this is a perishable gift existing for a brief period. In the moment, we can't imagine it will end. During these spells we are full of hope and dismissive of reason as our unbridled emotions lift the weight from everyday reality... These simple moments like some of those captured in Andy Warhol's intimate drawings of men from the 1950s remind us of the splendours life has to offer. A heart escaping from pursed lips, nude bodies sharing a

rapturous embrace, or a lustful stare are just a few examples of the simple, whimsical and affectionate moments Warhol captured in [these] drawings of love, sex and desire.[23]

Warhol's often tender, intimate line drawings of male lovers adorned with a tiny heart balanced on its pointed apex on the tip of an outstretched finger and liable to topple at any moment are 'imbued with an emotional vulnerability'.[24] They are perhaps all the more poignant coming from an artist who, in later work, so often chose to project himself and his 'Factory' as superficial and machine-like. Warhol biographer Blake Gopnik further illuminates the artist's pattern of work:

'Andy had this great passion for drawing people's cocks and he had pads and pads and pads of drawings of people's lower regions,' said a queer friend and assistant of Andy Warhol's. '…drawings of the penis, the balls and everything and there would be a little heart on them or tied with a little ribbon…'
Warhol did eventually get to show a sampling at least of his queer drawings in an exhibition called *Studies for a Boy Book*. The exhibition seems to have been mostly chaste portraits of Warhol's crushes, with maybe a few nudes with pubic hair and nipples romantically covered in hearts.[25, 26]

The drawings featured in a gallery exhibition that ran for two weeks, beginning, appropriately enough, on Valentine's Day 1956. The *New York Times*'s caustic, overtly homophobic review of the exhibition stated: 'Andy Warhol's figure drawings at the Bodley Gallery, 223 E 60th St, illustrate his "Book of Boy". They are sly, abound in private meaning and might have been done by Jean Cocteau on an off day.'[27]

Warhol draws on the heart amply; elsewhere, he showed trees laden with hearts as fruit in *A Tree to Make Love In*, as plumes of cigarette

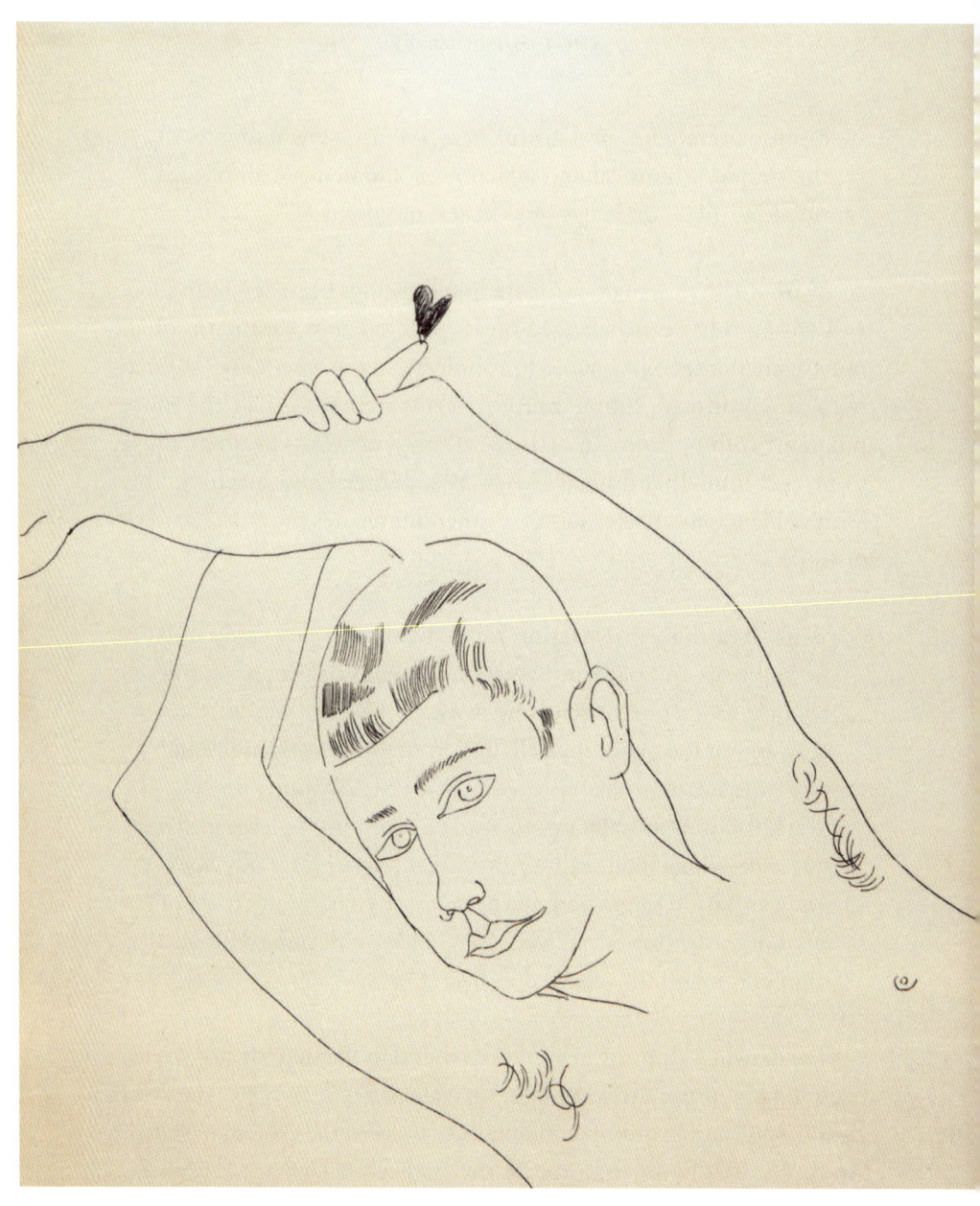

FIGURE 7.8 Andy Warhol, *Young Man with Heart*, c. 1954, New York.

smoke in *Boy with Hair on His Chest* or as the blooms of flowers, tattoos and simply as standalone motifs.

The heart depictions we have seen so far in this chapter vary from the deeply personal and expressive work of Frida Kahlo and Edvard Munch, whose hearts were drawn from a variety of complex and complementary sources to articulate particular and highly personal narratives, to the drawings of Andy Warhol, who used a more standard motif to communicate elements of what was then forbidden erotic love. But the broad question still remains why they and many others chose to depict the heart. What particular properties of the heart itself lead to its repeated translation into recognizable visual forms?

In earlier chapters, I have suggested that part of the reason the heart has been used in this way arises from the fact that it seems to exist apart from us – as a responsive element that both senses and transmits extreme emotional and physiological states. This idea of the heart having its own separate existence was demonstrated physiologically by Richard Lower, who showed that the explanted heart will continue to beat on its own for a short time, even after it has been removed from the body. In fact, there are experimental preparations in which the beating perfused heart can be maintained for many hours under optimally managed conditions. The same idea of autonomous (life-sustaining) function has been captured by the Norwegian writer Karl Ove Knausgård in the opening line of his novel *A Death in the Family*: 'For the heart, life is simple: it beats for as long as it can. Then it stops. Sooner or later, one day, this pounding action will cease of its own accord.'

I want to end the chapter by considering the work of Alexander Calder. Perhaps Calder's most significant contribution was to bring into art, specifically sculpture, a transition from three to four dimensions. He introduced the possibility of movement, of change in structure or position over time and, importantly, the realization that such changes can be brought about by interactions with the immediate environment. His mobiles were both self-contained and yet moved themselves in space, without mechanization. Movement, change over time and responsiveness to environmental cues are properties of the heart that, I have argued, are fundamental to the way we perceive it, which in turn influence the ways in which we portray it visually. The shared experiences that we have of the heart allow us to empathize with the artist's intent.

If anyone were to find a use for the responsive heart for artistic purposes, it should be Alexander Calder. I believe he intuited something of the properties of the heart, in a manner reminiscent of Aristotle when he observed the beating heart of the chick embryo and knew that this was the moving part that brought life.

Born in 1898, Calder attended the Art Students League in New York City from 1923–25, where he studied painting, etching, and drawing and prior to that had trained at the Stevens Institute of Technology, in Hoboken New Jersey, gaining a degree in Mechanical Engineering. He moved to Paris in 1926, where he befriended several avant-garde artists, including Fernand Léger, Jean Arp and Marcel Duchamp. Some of his early sculptural works were mechanized, with movement driven by motors, the moving elements comprising a rotating disc, arcing sphere or rotating helix, with no prescribed meaning but, to me, somewhat evocative of planetary systems. However, their movements were rigid, on a single axis and constrained by the mechanism. There was no scope for unpredictability, far less for spontaneity. Calder soon abandoned this type of model, instead drawing energy for the motion of his moving sculptures from environmental sources: the variable forces of air currents and human interventions.

In his biography of Alexander Calder, the art critic Jed Perl provides an account of the creation of Calder's kinetic sculptures, which Marcel Duchamp called 'mobiles'. Calder shared with the surrealists 'a sense that there were aspects of time and space – something mysterious and bewitching about the possibilities of the fourth dimension – that modern art had not yet adequately explored'.[28]

Perl notes that: 'Klee, Kandinsky and Miró had already made masterpieces out of arcs, spheres, densities and vectors, but in their paintings the movement was implicit, kinetic experience reconstructed in stillness… Calder's revolutionary mission was to trade the implicit for the explicit.'[29]

Writing in the avant-garde journal *Abstraction-Création: Art Non-Figuratif* in 1932, Calder himself noted:

> Nothing at all of this is fixed. Each element able to move, to stir, to oscillate, to come and go in its relationships with the other elements in its universe. It must not be just a 'fleeting' moment, but a physical bond between the varying events in life. Not extractions, but abstractions. Abstractions that are like nothing in life except in their manner of reacting.[30]

Most of the elements in Calder's mobiles were abstract shapes and forms, activated by unseen forces of nature. But he did incorporate a small number of figurative shapes, and among these the heart is prominent.

Part of his motivation in moving away from mechanized sculpture towards mobiles that were responsive to their environments – wind-powered rather than motorized – was captured in comments made to Calder by Piet Mondrian, who himself had written of dynamic aspects of art in relation to 'the conscious and the unconscious, the immutable and the mutable, emerging and changing shape under their reciprocal action'.

Jed Perl again draws attention to a feature that Calder had identified as important:

> When Duchamp had suggested that Calder called his new objects 'mobiles' both he and Calder were aware of the double meaning mobile has in French. As Calder later explained… 'Duchamp gave me the term he used for his own moving constructions

– "mobile". This in French means not only "movable" but also a "motive", a reason for an act, so I found it a very good word.'[31]

Duchamp himself had been experimenting with motion in painting. His work *Nude Descending a Staircase* (1912) is well known as an attempt to capture motion in a static painting. Here, a series of images of a woman in cubist style is effectively superimposed, capturing sequential poses of a woman descending stairs, in a format that we would now recognize as a kind of timelapse. Perhaps less well known is his *Coeurs Volants* (*Fluttering Hearts*) (Figure 7.9), in which a series of differently sized and subtly realigned alternating red and blue hearts is superimposed in a static simulation of the beating heart.

Calder had first used the simple heart motif in the 1920s, during a period when he was making figurative art fashioned from wires, and he would return to it in the jewellery he produced throughout the 1930s. From the 1947 piece *Untitled* (sheet metal, wire and paint) onwards, he also used the heart motif in his mobiles. He used the heart again in 1952 in *Valentine for Mary* (Figure 7.10) in which the hearts are the sole locomotive element (their movement comparable to the energy of wind captured by a sail), and he would come back to the motif several times right up until the 1970s.

I don't want to overstate the significance of the heart in Calder's work; it made up only a very small portion of the elements in his mobiles, and the actual motif itself is, in a sense, commonplace to the point of being trivial. Nonetheless, it seems to me significant that he uses the element repeatedly. The heart and its properties sit very comfortably with Calder's cherished notions of living art, automaticity, the sentience of the environment and mediating response.

In the 1946 text *Les mobiles de Calder*, Jean-Paul Sartre noted:

In the past, Calder drove them with an electric motor. Now he abandons them in the wild: in a garden by an open window he

FIGURE 7.9 Marcel Duchamp, *Coeurs Volants* (*Fluttering Hearts*), 1961, Stockholm.

lets them vibrate in the wind... they feed on the air, breathe it
and take their life from the indistinct life of the atmosphere.
[...] Calder's mobiles have a life of their own.[32]

So Calder was interested not only in movement but also in the
idea of his sculptures moving in response to their environment. As the
focus of his interests shifted towards random movement – shapes and
forms not under conscious control – it is no coincidence that among
the abstract cutout forms, he inserted heart motifs: the epitome of
unconscious movement. Indeed, this disconnect between the sensations
of the heart and the function of the conscious brain – the apparently
autonomous response of the heart to emotions and physiological
extremes – is fundamental to the way that we have come to regard it.

As the art historian Penelope Curtis writes in her essay 'Performance or Post-performance':

> Calder's distinction was in fact in allowing his sculpture to perform by itself. It is not time-limited, it does not shout or even narrate. Its presence is animate but softly modulated, speaking a language that no one shares, but which we sense. It is a kind of perpetual activity, which lies at the periphery of our sight, the boundaries of our senses…[33]

Sartre was struck by the way that, once crafted, the movement of Calder's mobiles was then dependent on the ever-changing character of natural forces.

> The forces at work are too numerous and complicated for any human mind, even that of their creator… For each of them Calder establishes a general fated course of movement, then abandons them to it: time, sun heat and wind will determine each particular dance. Thus the object is always midway between the servility of the statue and the independence of natural events. Each of its twists and turns is an inspiration of the moment.[34]

The playwright Arthur Miller made a similar point in his eulogy at Calder's memorial service in December 1976:

> It only slowly dawned on me that this work of cold wire and sheet metal was sensuous, that the ever-shifting relationships within a mobile were refracting the same elemental paradoxical forces in physics and human relations.[35]

Calder's sculptures are not depicting an event, they *are* the event. Or, more accurately, they are the stream of responses, not fully automatic or

FIGURE 7.10 Alexander Calder, *Valentine for Mary*, 1952.

self-contained in their movement, but sentient, perpetual, sometimes imperceptible and responsive – as is the heart.

Even this brief survey of the work of a small number of 20th-century artists suggests that the heart has retained a place as a modern icon. Its use sometimes seems intuited and at other times considered and deliberate. Though any 'classification' is unnecessarily constraining,

familiar themes and attributes surface. Pablo Picasso and Andy Warhol used the heart as a romantic (or erotic) emblem. In this context we might have looked at Miró's *Dancer* (1925) (Figure 7.11), in which the abstracted figure of the dancer on the right of the image has been portrayed with a moon-like sphere as her head, linked by a thin line to a voluptuous, crimson heart that dominates the figure, and which has symbolic genitalia attached directly to its apex. By using the heart in this way, Miró is clearly hinting at the erotic, at passion, at the notion of the heart as the core, but he is also associating it with movement – the heart is also the point of origin for the dancer's outstretched, arcing legs.

For Edvard Munch and Frida Kahlo, the heart was attached to intensity of emotion and, in particular, to grief, jealousy and loss. Salvador Dalí's *The Sacred Heart* (1929) is an ink drawing of the outline of Christ, or some saintly figure, with hand raised as if making the sign of the cross (Figure 7.12). In the chest is a small flaming heart surmounted by a cross and containing the words 'Sometimes I Spit with Pleasure on the Portrait of My Mother'.

Like Duchamp and Calder, Keith Haring and Henri Matisse used the heart as a focus of energy and the life force. Marcel Duchamp attempted to capture its movement and Alexander Calder, as we have seen, intuited something of the heart's automaticity and responsiveness.

While the strands are quite varied, the links with earlier depictions are striking and consistent. Significantly, even in this era – which was more biased towards personal expression – artists, for the most part, did not generate new heart types, either in their figurative representations (anatomical picture or motif) or in the repertoire of connotations.

I confess that when I first asked myself if modern artists, freed from the constraints of patronage and religious conformity to focus

FIGURE 7.11 Joan Miró, *Dancer*, 1925, Paris.

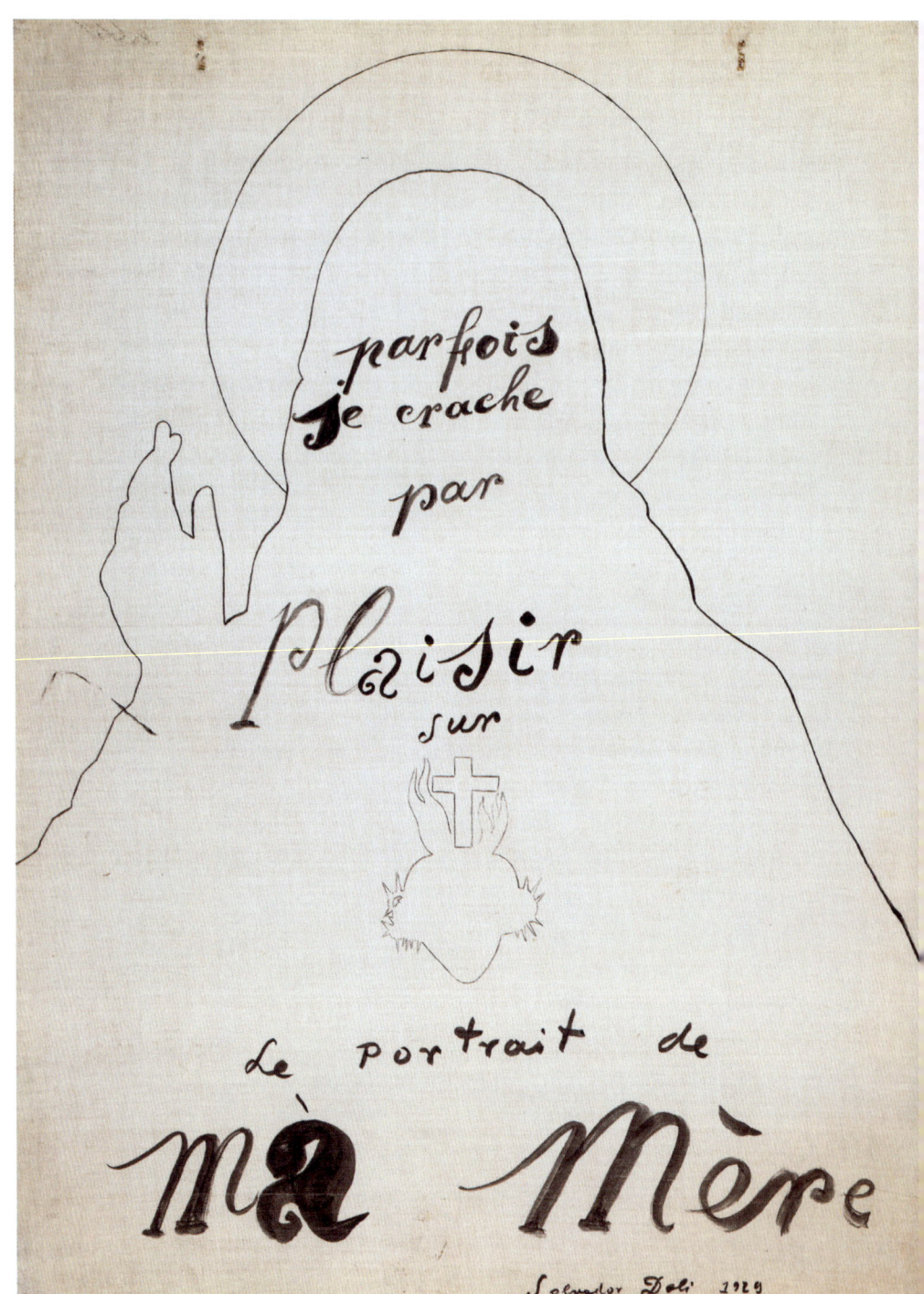

parfois
je crache
par

plaisir
sur

Le portrait de

ma Mère

Salvador Dalí 1929

on art as a means of personal expression, might draw on the heart in ways that echoed some of the motifs and devices from earlier times, I was only sketchily aware of a few examples. I approached the question with genuine curiosity, almost as if testing a scientific hypothesis. I had no sense of how pervasive the heart would turn out to be, nor of the number and diversity of artists, embracing varied styles and movements, who have captured something of the properties of this organ and its attributes to use as a device or vehicle in their own work. Although some have emerged, attempts to draw out particular strands of unbroken intellectual or cultural connectivity are not helpful.

A version of the heart is everywhere.

FIGURE 7.12 Salvador Dalí, *Sacred Heart*, 1929, Spain.

PACEMAKER FIBRE.   ZERO ANION CONDUCTANCE.

```
NA CONST   400       0.170    20     10.0     48     1.00
           0.14      0.10     42     15.0     0.12   82     5.0
K  CONST   1.20      0.0001   40     10.0     0.0020   80.0
           1.200     50    0.015     60     0.00
AN CONST   0    0.000   APPLIED CURRENT    0.000      10.000      0.(
```

T	V	M / NA-EFF	-INF	H / -GAIN	G-NA / K-EFF	N / -INF	G-K / -LOSS
0	-7.34	0.0409		0.8773	0.16	0.525	1.14
10	-7.56	0.21	20.38	2.09	41.31	21.40	-2.0(
20	-7.78	0.0419		0.8702	0.17	0.507	1.12
30	-7.99	0.21	20.51	6.28	40.89	20.85	-6.2
40	-8.20	0.0429		0.8630	0.17	0.491	1.11
50	-8.41	0.22	20.65	10.50	40.53	20.34	-10
60	-8.61	0.0439		0.8555	0.17	0.475	1.09
70	-8.82	0.22	20.80	14.75	40.20	19.87	-14
80	-9.02	0.0449		0.8480	0.17	0.460	1.07
90	-9.22	0.23	20.95	19.03	39.91	19.43	-18
100	-9.42	0.0459		0.8401	0.17	0.445	1.06
110	-9.62	0.23	21.11	23.35	39.65	19.02	-23
120	-9.82	0.0469		0.8320	0.17	0.431	1.05
130	-10.03	0.24	21.28	27.69	39.42	18.62	-27
140	-10.23	0.0480		0.8235	0.18	0.418	1.03
150	-10.44	0.24	21.47	32.07	39.21	18.24	-31
160	-10.65	0.0491		0.8144	0.18	0.406	1.02
170	-10.87	0.25	21.67	36.49	39.02	17.86	-36
180	-11.09	0.0503		0.8047	0.18	0.394	1.01
190	-11.32	0.26	21.90	40.96	38.85	17.48	-40
200	-11.56	0.0517		0.7939	0.18	0.382	1.00
210	-11.81	0.27	22.17	45.47	38.68	17.09	-44
220	-12.08	0.0532		0.7817	0.19	0.372	0.98
230	-12.37	0.27	22.49	50.04	38.52	16.67	-49
240	-12.68	0.0550		0.7672	0.19	0.361	0.97
250	-13.03	0.29	23.90	54.69	38.36	16.19	-53
260	-13.42	0.0572		0.7489	0.20	0.352	0.95
270	-13.88	0.30	23.48	59.43	38.19	15.61	-58
280	-14.44	0.0605		0.7236	0.20	0.343	0.93
290	-15.13	0.33	24.45	64.33	37.98	14.82	-63
300	-16.07	0.0661		0.6827	0.22	0.335	0.90
310	-17.46	0.39	26.68	69.52	37.65	13.46	-68
320	-19.89	0.0810		0.5912	0.27	0.328	0.84
330	-26.15	0.87	42.31	75.93	36.45	9.41	-73

The Beating Heart

If we suppose that the heart beats the way Harvey describes it, we would have to imagine some faculty causes this motion, and the nature of this faculty would be much more difficult to understand than what it claims to explain.

RENÉ DESCARTES
La Description du Corps Humain, 1664

In the cardiac cycle, diastole is the period of active relaxation. Diastole occurs between the contractions; it is the longer part of the cycle, and it has a slower, almost deliberative quality; it is the period during which the heart fills. Indeed, the length of the filling phase and the volume of blood accepted by the heart in part determine the force of the next contraction. Although commonly used, the concept of relaxation of the heart in diastole is somewhat misleading, since the molecular and cellular processes required to effect it are far from passive and require both complex coordinated activity and the expenditure of cellular energy.

The prolific 'scientific' outputs of the sixteenth and seventeenth centuries in relation to the function of the heart and circulation were, to some extent, followed by a pause. That is not to say that nothing was happening; the world was clearly changing through the eighteenth and – especially – the nineteenth centuries. Through industrialization, new technologies and the refinement and codification of the scientific method, the stage was being set for the quantum leaps of twentieth-century science and medicine – but important new findings in relation to the heart did not emerge in this period: there was a diastole.

We can recall from earlier chapters those long periods in which the theories and practice of Western medicine were simply handed on from generation to generation, where the mark of scholarship was faithful repetition rather than original observation. Reconciliation of many previously inconsistent ideas and observations had occurred by the end of the seventeenth century and, as far as the heart was concerned, by the end of that period the Galenic synthesis had been laid to rest. The early physiologists trying to understand the workings of the heart no longer believed that blood was made by the liver, or that it crossed the interventricular septum through imaginary pores, to be vivified by mixing with air from the lungs. They knew that it did not ebb slowly to be consumed by the tissues. William Harvey had built a theory of circulation based on his observations that the veins allowed only one-way passage of blood. He allied that theory with quantitative reasoning that led him to conclude that the liver could never make blood fast enough to keep up with the rate of its passage through the arteries, and that the blood must circulate in a loop. Richard Lower followed Harvey by showing how the heart could continue to beat even after the blood of a dog had been replaced with beer, and he observed that even after it had been removed from the body, the explanted heart kept beating for a time, thereby reaffirming the apparent autonomy of this unique organ.

But one critical mystery remained. By what force or power did the heart bring itself to beat? The lack of an explanation for the automaticity of the heart had puzzled Thomas Aquinas and was one of René Descartes's principal objections to Harvey's interpretation of the heart as a pump to effect the circulation of blood. Despite his inclination to view the world in mechanical terms, Descartes did not fully accept Harvey's notion that the heart was simply a pump that propelled blood through the body. In part, this reflected Descartes's belief that all muscles were controlled by the will. In *La Description du Corps Humain* (*The Description of the Human Body*, published posthumously in 1664), he wrote:

> If we suppose that the heart beats the way Harvey describes it, we would have to imagine some faculty causes this motion, and the nature of this faculty would be much more difficult to understand than what it claims to explain.[1]

Richard Lower went further in his *Tractatus de Corde* (1669):

> I should speak here of the ultimate way in which the heart's movement is effected, but as it is over difficult to obtain any true conception of this and it is the privilege of God alone, who comprehends the heart's secrets, to understand its movement also, I will not waste effort in examining it further.[2]

Lower was right: the solution was completely inaccessible to the science of his era. The tools and techniques required to understand how the heart instigated its own beat would not become available for almost three hundred years.

The period of diastole through the eighteenth and early nineteenth centuries yielded only a few incremental gains in knowledge of heart structure and function. Nonetheless, by the nineteenth and twentieth centuries, the biological world had come to be regarded on an entirely different scale; the microscope had laid open the possibility of conceiving structures at cellular and even subcellular levels. Accordingly, the unit of consideration was no longer the organism or the organ but the individual cell.

Significantly, the influential work of physiologists such as Claude Bernard (whom we encountered in Chapter 7) had made a profound impression and laid open possibilities to engage more deeply with fundamental questions. One of Bernard's crucial contributions was to pull the biological sciences into the realm of physical and chemical laws and, as a result, to provide impetus in the field to apply and extend systematic experimentation and mathematical modelling in the biological systems.

In Chapter 4, I mentioned Dr Jonathan Miller's theory that Harvey's notion of the heart as a pump might have been prompted by the growing number of mechanical tools being developed in the early seventeenth-century world he inhabited. By analogy, early twentieth-century investigators were exploring new concepts in the fields of electric charge and current and had devised the tools to measure them, including at a microscopic scale. By understanding biological tissues at a physical and chemical level, it was possible to imagine altogether new possibilities for their function. From the 1930s, a concept of 'excitable tissue' began to emerge.

Measuring the electrical activity within individual nerves, given their extremely small size and delicate structure, is exceptionally challenging.

By identifying the squid giant axon – which is about ten times the diameter of a human nerve cell – researchers had become able to study neuron (nerve cell) function by inserting tiny hand-pulled glass tubes (microelectrodes) that would allow them to measure the electrical charge across the cell membrane (the insulating lipid boundary that divides the inside of the cell from the outside of the cell).* In essence, when a neuron is at rest, a variety of charged particles, called ions, are asymmetrically distributed between the inside and the outside of the cell. In the basal state (non electrically-excited), positively charged sodium ions are maintained outside the neuron, while positively charged potassium ions are actively sequestered on the inside. The asymmetric distribution of these charged particles results in so-called membrane potential, in which the inside of the neuron is negatively charged with respect to the outside.

During an 'action potential', the properties of the insulating cell membrane change and allow movement of positively charged sodium ions from the outside to the inside through small gates known as ion channels. Importantly, different classes of ion channel have their own characteristics (in terms of what they allow to pass and when) and are inherently sensitive to the magnitude of the potential (i.e. the voltage) across the membrane. The permeability of the voltage-gated ion channels changes in response to the potential across the membrane, which in turn is affected by the flow of ions they permit. Changes in the flux of ions result in a burst of electrical activity that can be detected using the glass microelectrodes and forms the basis for the propagation of information along the length of the neuron. The combination of experimental observations and mathematical modelling that allowed these processes

---

* The squid giant axon is an unusually large 'limb' of a nerve cell that extends along the body of a squid to effect rapid nerve conduction and ultimately activates propulsion. It was described in the early part of the twentieth century and its significance was immortalized as the cell from which the mechanism of electrical conduction of nerve cells was elucidated.

to be understood led to Alan Hodgkin and Andrew Huxley being awarded the Nobel Prize for Medicine or Physiology in 1963. They had uncovered the key to the propagation of neuronal electrical impulses underlying the function of the nervous system.

The discovery of an analogous electrical phenomenon in the heart is credited to the Swiss-born Silvio Weidmann. Weidmann arrived in Cambridge during the summer of 1948, shortly after Alan Hodgkin returned from a visit to the University of Chicago, where he had learned to make the new fine-tipped microelectrodes that formed the basis for his work with Andrew Huxley on the ionic basis of the nerve action potential.

On 16 July 1949 in Cambridge, during afternoon tea, Weidmann was offered the opportunity to work on the heart of a dog that had been used for a student demonstration of Starling's law (the phenomenon whereby the force of contraction of the heart is related to volume of blood it receives). Working into the night and using the new microelectrode technique, he recorded the first quantitatively accurate action potential derived not from a neuron but from a cell of the heart itself: a cardiac action potential. In October 1949, he and Edouard Coraboeuf published an account of this work, just before others reported measurements of the frog cardiac action potential.[3, 4]

In other words, the preliminary investigation of heart tissue had revealed that it was 'excitable'. The presence of an action potential indicated that it was an electrically active organ. However, unlike the nerve cell, which receives an external stimulus to initiate the action potential, the action potential in the heart occurred rhythmically, repetitively and spontaneously. This was essentially a restatement, in twentieth-century terms, of the challenge articulated by Descartes and Lower some three hundred years earlier. To uncover the cellular and molecular basis for this phenomenon would be to unpick what Aristotle had identified as the origin of life.

Listening to Professor Denis Noble recall, with great modesty, the circumstances and setting in which he came to address and unravel this age-old problem has been the highlight of my research for this book. In truth, I already knew the story from the perspective of a cardiologist: how the ion channels work and how they might be susceptible to this drug or that. I had heard Noble relate a version of the story before and I had read his own account in his book *The Music of Life*.[5] But I had not thought about his discovery in its historical context or stopped to think about his personal experience of unpicking this elusive problem.

Noble was drawn to this hitherto intractable challenge when he was a graduate student at University College London (UCL). His imagination had been captured by the work of the group in Cambridge. He recalls a conversation with the German physiologist and physician Rolf Niedergerke, who had discovered that muscle contraction comes about through shortening of the muscle fibres (the so-called sliding filament model), and who told Noble: 'of course the big problem is that nobody knows why it beats'. By the time of that conversation, Noble had, in fact, worked out exactly how the heart generated its own beat, but (in a manner reminiscent of William Harvey's nine years of reticence) did not have the confidence to announce the fact to the more senior scientist. Very few research groups around the world had been addressing this problem, in large part, as Noble recalls, because the experimentation was so 'fiendishly difficult'. Even as Noble began his work on the heart cell, his doctoral supervisor at UCL, Otto Hutter, in conjunction with Wolfgang Trautwein in Baltimore, had already demonstrated that the rate of the spontaneous electrical discharge of cells in the pacemaker region of the heart (the sinus node) was influenced by the autonomic nervous system – a part of the nervous system that responds to physiological and emotional conditions but is not under voluntary control. Specifically, sympathetic stimulation sped the pacemaker rate up, while vagal (parasympathetic stimulation) slowed it down.[6] These nerves are the very same that Richard Lower had so carefully depicted in Thomas Willis's *Cerebri Anatome* (Figure 5.6) without any knowledge of their function. In fact, it would eventually transpire that both the rate and force of contraction of the heart are influenced by their outputs.

Working towards his doctoral thesis, under the supervision of Hutter, Noble had been measuring the action potential in the Purkinje fibres of sheep hearts, which he himself procured fresh through early morning visits to an abattoir in north London. By day, he was devoted to the experimental work. He undertook the intricate and laborious process of fashioning, in a Bunsen burner flame, the glass microelectrodes (maybe one in a hundred actually worked) that he could use to penetrate the membranes of the painstakingly isolated Purkinje fibres. Having manufactured the microelectrodes and isolated and nurtured the isolated cells, he could start the intricate task of experimentally determining their electrical behaviour under various conditions. Close to his laboratory at UCL was London's emerging centre for the purchase of electronic goods, on Tottenham Court Road. Here Noble procured the specialist ME1400 valves that provided sufficiently high input impedance for him to draw tiny currents without destroying the precarious cell preparations. Everything was done by hand.

Noble was particularly influenced by the work of Hodgkin and Huxley, which had allowed the neuronal action potential to be modelled with mathematical equations. He reasoned that an equivalent approach ought to be possible in the heart, but at that time no one in that field was undertaking such work. The calculations were complex and to solve the problem in any practical way would require computation. Therein lay a problem. In the late 1950s, not only was computation in its infancy, but there were almost no accessible computers. Noble approached the group responsible for the access and operation of Mercury, then the only computer at UCL, but no one expected a biologist to show interest. The computer was, Noble recalls, the 'preserve of computer scientists, particle physicists, and astronomers'. His request was initially declined, but he must have made an impressive case, since he was eventually granted access. However, he was allowed to use Mercury only between 2 a.m. and 4 a.m. At the outset, he sketched the essence of the problem, hoping to secure collaboration and support from a programmer, but not only was this the only computer in the university, the expertise to interact with it was sparse. It is worth pausing to reflect that this was not a computer that functioned in the manner of computers we are familiar with today. There was no convenient software application, no out-of-the-box code, not even a screen. There was virtually no usable interface. The input was punched tape; the output was tape.

Smiling as he relates to me the nature of the obstacles, Noble recalls being told: 'Well, Mr Noble, we don't really think you understand what you will have to do – you will have to write all the code', and furthermore, 'we have looked at your equations and there is no oscillator there'. Even the computational scientists were stating, in mathematical terms, the same problem that had been posed by Descartes. Where was the external driver of rhythm?

In Noble's recollection for *The Music of Life*, he writes:

> Yet, it looks like a reasonable question. In a system that oscillates it seems that there must be some specific component that oscillates, around which the behaviour of the entire system is geared; and there must be a mathematical function that describes the way that the component oscillates. Indeed, it is an eminently necessary question, if we are talking about some man made mechanical systems. But we are not. Instead, we can have a system that operates rhythmically and yet contains no specific 'oscillator' component.[7]

In essence, Noble's approach was to undertake the experimental work that allowed him to observe what happened at the level of the cell and to determine the properties of the various critical components. In the complementary computational work, he then varied the conditions to simulate the action potential and to try and understand how the cell system worked – and specifically how the changes in current and electrical potential might lead to spontaneous action potential generation that could underlie the heart's ability to beat on its own.

One of the obstacles to progress was the extreme slowness of the computation. Trial and error of the various parameters could generate implausible biological simulations, but the processing was so slow and opaque that any modelling failure could take hours to reveal itself. To increase the efficiency of his system, Noble programmed in parameter limits that, when breached, triggered the computer to output the tune 'Oh Dear! What Can the Matter Be?' through its primitive speaker. The full-time computational scientists, on hearing this apparently frivolous use of the machine, may have doubted the seriousness of this nocturnal Mercury user, but Noble was able to use the tune to increase the throughput of experimental parameters, eventually culminating in a set that worked and stimulated the measured pacemaker potential.

Noble recalls: 'I put in a tape that had the tweaking of parameters that actually gave the pacemaker depolarisation and I thought… if I play with this a bit more won't it take off automatically?'

While the graphic image (Figure 8.3) of the action potential is easily visualized now, Mercury generated data that emerged first as paper tape that Noble had to feed into a teleprinter. This spewed out vast columns of tabulated numbers. Noble remembers the data:

> pouring out and eventually of course [I] became reasonably good at seeing whether it was working, just by eyeballing how the voltage was changing and that's when I knew of course that, yes, there was a clear depolarization occurring… slow depolarization… that excited me obviously.
>
> …it was only later when I got the graph paper out and started plotting that I got even more excited because this really did look not only like a pacemaker depolarization but roughly the right time

course… and there it was… reaching up towards the threshold. …and once I got that I was over the moon and that's where I had got to when I had the conversation with Rolf Niedergerke over tea… Given how high up he was over me I did not dare tell him that.

But Noble did tell Otto Hutter, who responded (positively, albeit with jaw-dropping understatement): 'Oh my goodness, you've got your thesis.'

Figure 8.2 shows the tabulated data that Noble presented for his 1961 PhD thesis. I asked him if I could see and photograph the pages in question, and he graciously obliged. Here are the data that underpinned the description of the initiation of the heartbeat. When Noble handed me his original thesis in the corner of the Senior Common Room in Balliol, it was a sunny afternoon and we had arranged to meet for lunch. I had arrived from the hospital slightly late and was pretending to be less agitated than I was at my rudeness in keeping him waiting. I had not imagined that he would actually bring, let alone part with, the original thesis. I could not have been more in awe of the document in my hand if it had been a Gutenberg Bible or the Magna Carta.

Shortly before his twenty-fourth birthday, when he was still a graduate student, Noble's findings were published as a letter to *Nature*, in which, remarkably, the young man was the sole author.[8] Characteristically for both the man and perhaps the era, the presentation is crisp, to the point and, given the gravity of the findings, wholly understated (the emphasis is mine):

In these equations there is always at least one potential at which the steady state sodium and potassium currents are equal and opposite. In the computations just described, this forms the resting potential and the system is stable unless excited. *A small change in the constants in the equation's form is sufficient to make*

180

Pacemaker solutions.

Two solutions are shown. The first (pp 181-183) is the solution with zero anion conductance and is the result shown in Fig. 19A. The fluxes were also computed and these are plotted in Fig. 21C. The result shown on pages 184 & 185 shows the effect of adding an anion conductance of $0 \cdot 18$ mmho/cm^2 with E_{An} = -60 mV. This is plotted in Fig. 30C.

FIGURE 8.1 Pacemaker Solutions; reproduced from Noble, D., PhD thesis, *Ion Conductance of Cardiac Muscle*, 1961, University College, London. With permission of the author.

```
PACEMAKER FIBRE.        ZERO ANION CONDUCTANCE.

NA CONST   400     0.170    20    10.0    48    1.00
           0.14    0.10     42    15.0    0.12   82    5.0
K  CONST   1.20    0.0001   40    10.0    0.0020  80.0
           1.200   50    0.015    60    0.00
AN CONST   0    0.000   APPLIED CURRENT   0.000    10.000    0.00
```

T	V	M / NA-EFF	H / -INF	G-NA / -GAIN	N / K-EFF	-INF	G-K / -LOSS
0	-7.34	0.0409	0.8773	0.16	0.525		1.14
10	-7.56	0.21	20.38	2.09	41.31	21.40	-2.06
20	-7.78	0.0419	0.8702	0.17	0.507		1.12
30	-7.99	0.21	20.51	6.28	40.89	20.85	-6.20
40	-8.20	0.0429	0.8630	0.17	0.491		1.11
50	-8.41	0.22	20.65	10.50	40.53	20.34	-10.37
60	-8.61	0.0439	0.8555	0.17	0.475		1.09
70	-8.82	0.22	20.80	14.75	40.20	19.87	-14.57
80	-9.02	0.0449	0.8480	0.17	0.460		1.07
90	-9.22	0.23	20.95	19.03	39.91	19.43	-18.80
100	-9.42	0.0459	0.8401	0.17	0.445		1.06
110	-9.62	0.23	21.11	23.35	39.65	19.02	-23.06
120	-9.82	0.0469	0.8320	0.17	0.431		1.05
130	-10.03	0.24	21.28	27.69	39.42	18.62	-27.36
140	-10.23	0.0480	0.8235	0.18	0.418		1.03
150	-10.44	0.24	21.47	32.07	39.21	18.24	-31.69
160	-10.65	0.0491	0.8144	0.18	0.406		1.02
170	-10.87	0.25	21.67	36.49	39.02	17.86	-36.05
180	-11.09	0.0503	0.8047	0.18	0.394		1.01
190	-11.32	0.26	21.90	40.96	38.85	17.48	-40.46
200	-11.56	0.0517	0.7939	0.18	0.382		1.00
210	-11.81	0.27	22.17	45.47	38.68	17.09	-44.91
220	-12.08	0.0532	0.7817	0.19	0.372		0.98
230	-12.37	0.27	22.49	50.04	38.52	16.67	-49.42
240	-12.68	0.0550	0.7672	0.19	0.361		0.97
250	-13.03	0.29	22.90	54.69	38.36	16.19	-53.98
260	-13.42	0.0572	0.7489	0.20	0.352		0.95
270	-13.88	0.30	23.48	59.43	38.19	15.61	-58.62
280	-14.44	0.0605	0.7236	0.20	0.343		0.93
290	-15.13	0.33	24.45	64.33	37.98	14.82	-63.36
300	-16.07	0.0661	0.6827	0.22	0.335		0.90
310	-17.46	0.39	26.68	69.52	37.65	13.46	-68.26
320	-19.89	0.0810	0.5912	0.27	0.328		0.84
330	-26.15	0.87	42.31	75.93	36.45	9.41	-73.51 →

FIGURE 8.2 Conditions for the spontaneous pacemaker activity of the heart; reproduced from Noble, D., PhD thesis, *Ion Conductance of Cardiac Muscle*, 1961, University College, London. With permission of the author.

the system unstable in diastole, and pacemaker activity then occurs. Such a solution is shown in Figure 2.

Noble had modelled the conditions and ion channel properties that would allow the heart to cycle the electrical activity autonomously and repeatedly, thereby ensuring the rhythmic electrical activity that underlies heart muscle contraction and consequently an automatic heartbeat. The demonstration of the action potential solution is shown in Figure 3 from the letter to *Nature* (Figure 8.3). The 'pacemaker activity' refers to the gently upsloping contour that precedes each of the near-vertical up spikes and represents a slowly timed creep of the membrane potential up to 'threshold potential', triggering opening of the voltage-gated channels and firing the action potential. Subsequent closure of these channels and active ion pumping would lead to restitution of the membrane potential and the whole process would start again. In other words, the right conditions would lead to spontaneous, repetitive electrical activity in the heart.

In the context of the developing embryo, stirring into visible life, the cells of the developing heart need only form themselves, with their insulating outer membrane, synthesize and orientate the necessary ion channels in that membrane and, subject to physiological conditions in the environment in which it is bathed, a heart cell will start to beat, unassisted. Noble's solution would explain the rhythmic contraction that Aristotle had seen through the window in a fertilized chick egg and identified as the origin of life.

Pushed on whether he realized the significance of his discovery at the time, Noble reflected:

I never imagined that something as central to life as finding out how the rhythm generator of the heart would work should fall into my lap like that.

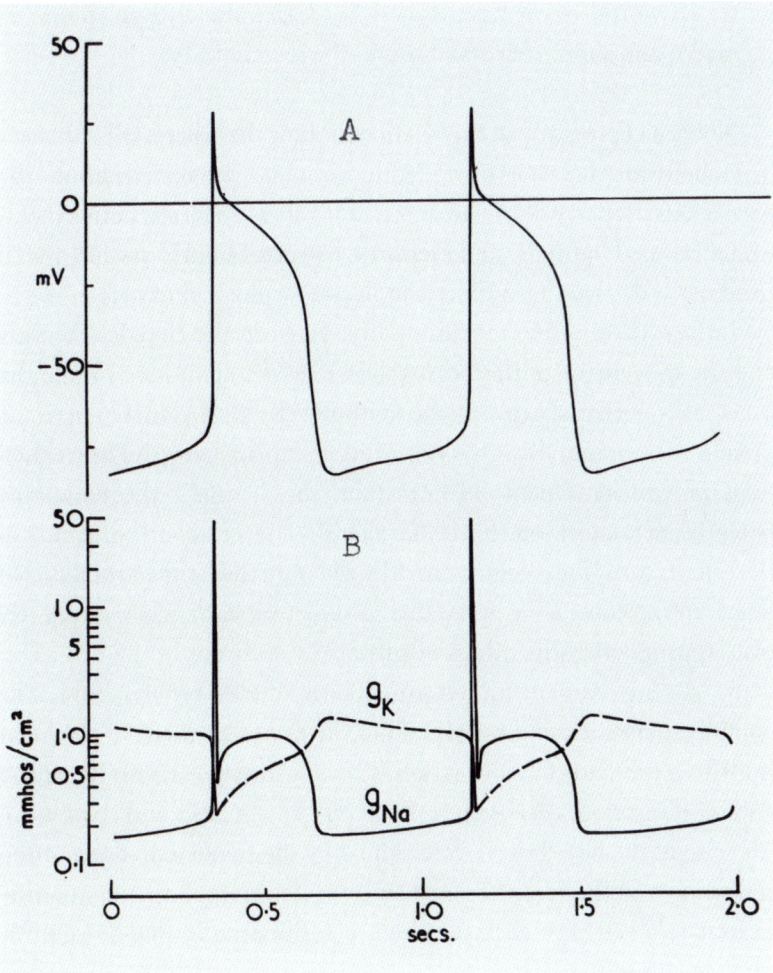

FIGURE 8.3 Computed action and pacemaker potentials. Figure 19 from Noble, D., PhD thesis, *Ion Conductance of Cardiac Muscle*, 1961, University College, London. With permission of the author.

[...] I would never have done what Crick did and go into a tavern and shout that I had found the secret of life.*

Noble had provided an explanation for how the heart is self-sufficient in its ability to generate its own rhythm without external stimulation. The mysterious property of the heart – that it generates its own movement, which Lower, Highmore and Descartes had attributed to the life-giving hand of God – had been described in purely physical terms.

In fact, there is one remaining link between the electrical activity and the movement of the heart that we need to consider. While the pacemaker potential explains the rhythmic electrical activity across the plasma membrane, this does not of itself explain how the heart then responds with a mechanical contraction. The changes in the membrane potential have knock-on effects that alter the concentration of calcium in the cell. In turn, the calcium provides the impetus for the proteins that make up the contractile apparatus to slide over each other, effectively overlapping, with the effect of physically shortening the cell. This is the 'sliding filament model' to which I alluded briefly above. The combined effect extending over all of the muscular tissue of the heart leads to a mechanical contraction. Cycling of the calcium, its release and sequestration, means that the heart can contract and then relax. The rate of the heartbeat is determined by the frequency of the action potential, and the force of contraction is determined by the quantity of available calcium, each of which is influenced by the autonomic nervous system.

These properties, in turn, form the basis for the heart seeming to function independently, beyond but perceptible to the conscious self. Its rhythm, rate and force of contraction are all controlled in response

---

* In James Watson's account, after the discovery of the structure of the double helix (which enables DNA to carry and replicate the genetic code), Francis Crick burst into The Eagle pub in Cambridge and announced that he and Watson had discovered the secret of life.[9]

to complex stimuli by processes that are seemingly physiologically autonomous. The intermittent awareness of change in the heart in these respects forms the basis of our experience that the heart is something with its own properties and, in a sense, its own sentient existence.

While able to generate its own rhythm, the heart is not completely disconnected from the body it inhabits. Some connectivity to the body is important both for determining its precise behaviours and for transmitting a sense of its status to the body. The heart has two sets of neuronal connections; each is part of what is now termed the autonomic nervous system.[10] The distinction of this part of the nervous system is that, while it is connected to the brain, it does not enable or permit direct voluntary inputs to the organ(s) it supplies. This is the part of the nervous system that mediates some core functional reflexes, such as managing constant blood pressure as we move from lying to standing, or in driving the increased heart rate and pump output that accompany exercise. Lower described the innervation of the heart by the vagus nerve, which is central to these processes. The vagus nerve calms the heart; it suppresses the heart rate and the force of contraction. The vagus nerve is also important because it carries sensory inputs from the heart to the brain. In other words, the vagus nerve is important in the processes of interoception: our awareness of internal bodily status. I have made the claim repeatedly that the combination of awareness of the heart alongside its involuntary (and partly autonomous) responsiveness accounts for the ways in which we perceive and portray it, and that our shared experiences of the heart make it a meaningful tool for communication. We all understand the language of the heart, based on our individual experiences of our own hearts, which are at the same time held in common.

The other class of nerves reaching the heart comes from the brain via the spinal cord and thoracic ganglions. This set of nerves provides a broadly opposing function to the vagus nerve. Stimulation of the so-called sympathetic nerves increases the heart rate and the force of contraction and accelerates the conduction of the electrical impulse through the heart muscle itself.

For now, I simply want to emphasize that while the heart beats on its own, its force of contraction and rate of beating are influenced by the autonomic nervous system, which integrates complex information and regulates heart function in ways that are almost entirely outside our voluntary control. Importantly, the vagus nerve takes sensory information to the brain so that we are aware of some of these changes.

Knowing that the vagus nerve takes information from the heart to the brain is one thing, but it is not quite the same as knowing that the way we feel and experience the heart is mediated in this way. For instance, the vagus nerve could transmit physiological data back to the heart, while the sensation of the heartbeat could, in theory, be transmitted by the somatosensory nerves, such as those that innervate the chest wall. In theory, we might be aware of our heart beating only to the extent that it pounds on adjacent structures, such as the chest wall. We know this is not the case through observations of the effects of surgically removing the nerves to the human heart. During heart transplantation, the nerves to the heart itself are severed. When the donor heart is implanted, although the 'plumbing' is reconnected, the nerves are not, so the heart is left, from a neurological point of view, disconnected and freestanding. As a result, the neuronal modulation of the heart's function is removed and the perception of the heart is blunted.[11]

The nervous attachments of the heart also seem to be responsible for 'takotsubo cardiomyopathy' or 'broken heart syndrome'. It has long been anecdotally recognized that from time to time a person under extreme physical or emotional stress (and particularly bereavement) can suffer symptoms suggestive of a heart attack that can even lead to death. We now recognize a condition in which part of the heart muscle becomes transiently stunned and completely inactive. As a result, when the healthy part of the heart muscle at its base constricts and the paralysed apex expands outwards, the shape resembles a Japanese octopus pot, or *takotsubo*. The appearances are very characteristic. About one-quarter of patients with this condition have experienced an identifiable recent emotional stress, of which bereavement is probably the most common.[12] The molecular and cellular causes are not completely understood, but the relationship with emotional or physical stress and the circumferential distribution of the heart muscle paralysis suggest that activation of the sympathetic portion of the autonomic nervous system or circulating adrenaline-type compounds are likely to be important drivers. For most patients the progression is benign, with restoration of normal function over a period of weeks. However, in a significant minority, a range of complications can occur, including death. It really is possible to die from a broken heart.

So there we have it. The heart is a pump in a circulatory system that generates its own rhythmic contraction through the cyclical movement of metal ions through voltage-gated channels, the properties of which are, in turn, influenced by nervous inputs that change the heart rate and the force of contraction. The initiation and propagation of the heartbeat can be modelled by computer and described in columns of numbers – the enthralling magic of its automaticity and responsiveness reduced to lines on a graph. The mystery that so captivated thinkers from Aristotle and Aquinas to Leonardo, Descartes and Harvey has been comprehensively solved. The mystery has gone. There is surely no ghost in this machine.

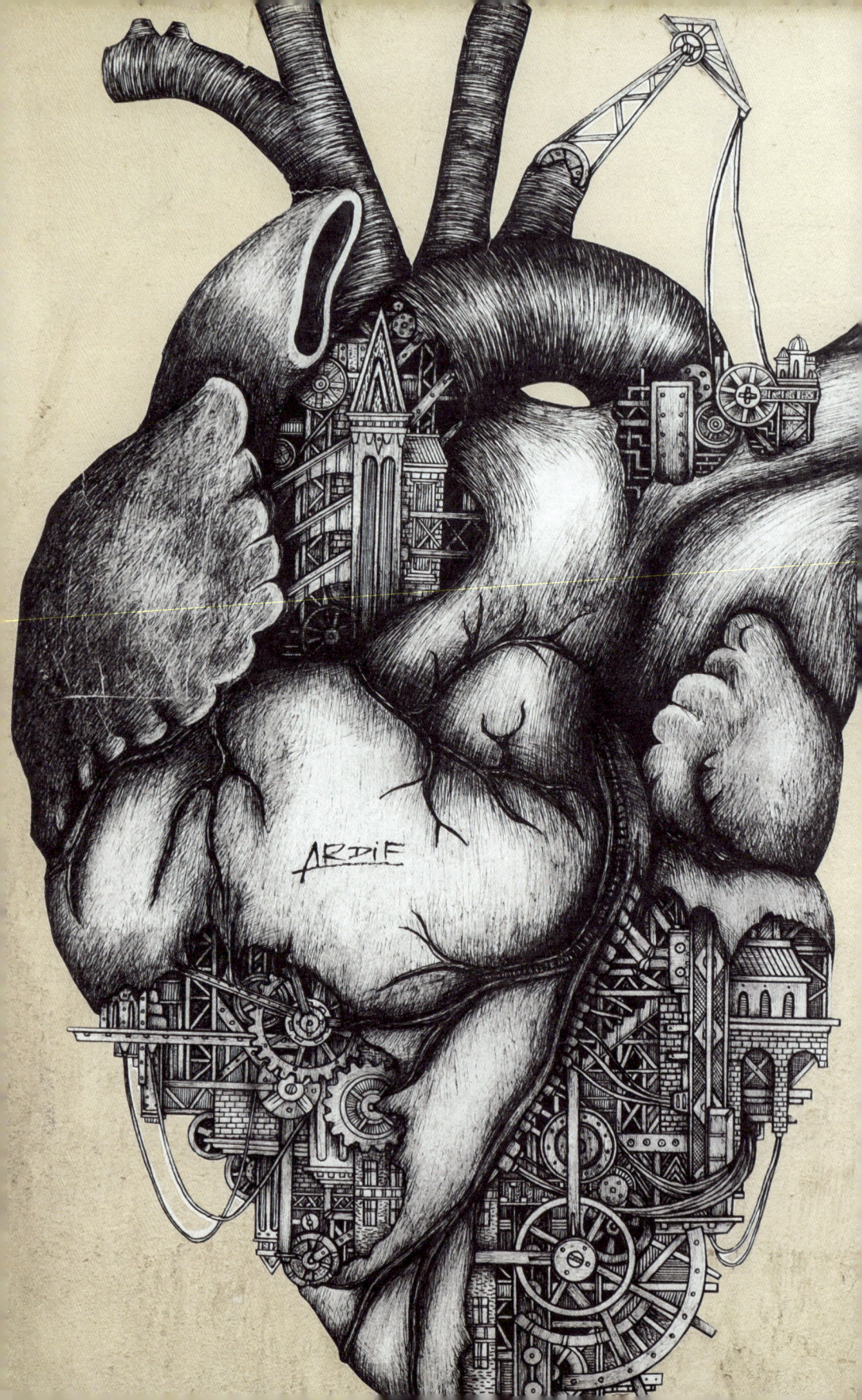

9

The Contemporary Heart: A Ghost in the Machine

It's a fool that looks for logic in the chambers of the human heart.

JOEL COEN
O Brother, Where Art Thou?, 2000

When I embarked on this exploration, I had hoped to attempt to explain – but certainly not to explain *away* – the remarkably consistent attributes that have been conferred on the heart and expressed in images of it. I have tried to understand how beliefs about the heart have changed over a period of more than two thousand years, reflecting the cultural, spiritual, intellectual and, eventually, scientific contexts of the time. Often armed with only partial knowledge of what needed to be explained, humanity has deployed the heart to plug gaps in its understanding; it has been the dwelling place of the soul, the source of life, a furnace or fermenter providing the heat of living bodies, the source of semen and a repository of deeds. It has become the seat of love and desire.

I am going to resist any temptation to imply that we now understand all there is to know about the heart's function, its physiological responses or how we perceive them; there will be more discoveries to come. But after all of that, even contemporary knowledge tells us that the heart is a pump to circulate the blood, generating its own contraction through the prosaic inevitability of the passage of ions obeying rigid physicochemical laws.

And yet, our language, idiom and visual world still contain abundant references to the heart and its imagined properties. If you doubt this, stop to take a look at the enormous range of heart motifs that feature in contemporary street art. These are often expressive pieces and, for the most part, one imagines that they don't make formal reference to any canon of art. In that sense, they are 'from the heart'. You can hardly take more than a few paces in the streets of London, Lisbon, Paris, Phnom Penh, Pondicherry, New York, Naples, Rome, Cairo or Marrakech without stumbling across the heart motif in one form or another. Heart images adorn walls, letter boxes, streetlamps, phone kiosks and, not infrequently, the ground. They dance around windows and climb the frames of doors. Over the years, I have collected several hundred photographs of these hearts. Of course, we can all bring to mind trite examples of roughly scribbled or coarsely scratched hearts, punctured by an arrow and flanked by a couple of unruly scrawled initials, marking some ephemeral juvenile love affair. But there are many examples of more nuanced and sometimes very poignant uses of the heart as expressions of love, grief, injury and outrage, or for some serious (or not so serious) political statement.

After the tragic 2017 fire at Grenfell Tower in London's North Kensington, walls of the adjacent Notting Hill streets saw a proliferation of eponymous green hearts, shown in various forms, most of which seemed intended not only to convey profound grief and loss but to perfuse the area with a living memorial to the victims. Some were also designed to transmit visceral outrage at the neglect and corruption that

FIGURE 9.1 (above) Grenfell Tower; community memorial under the Westway, 2022, London.

FIGURE 9.2 (overleaf) Contemporary street art images: New York; Bologna; London; Granada; Madrid; Florence; Pondicherry; Paris.

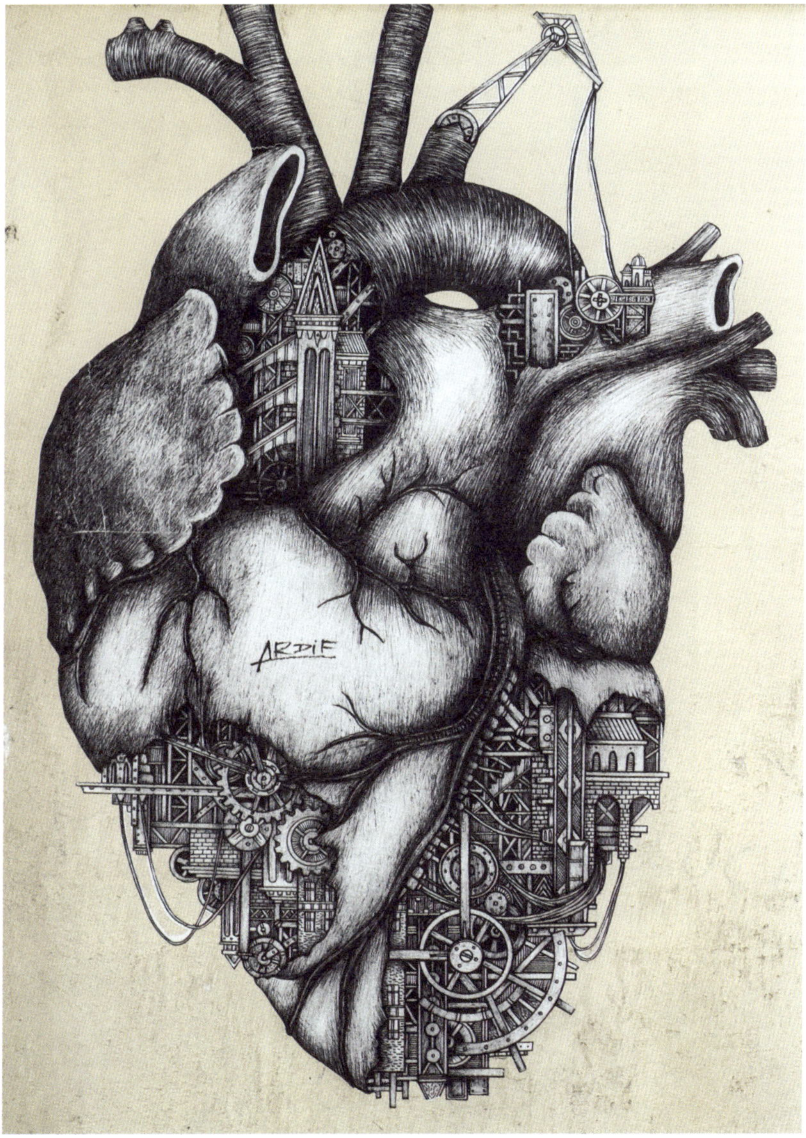

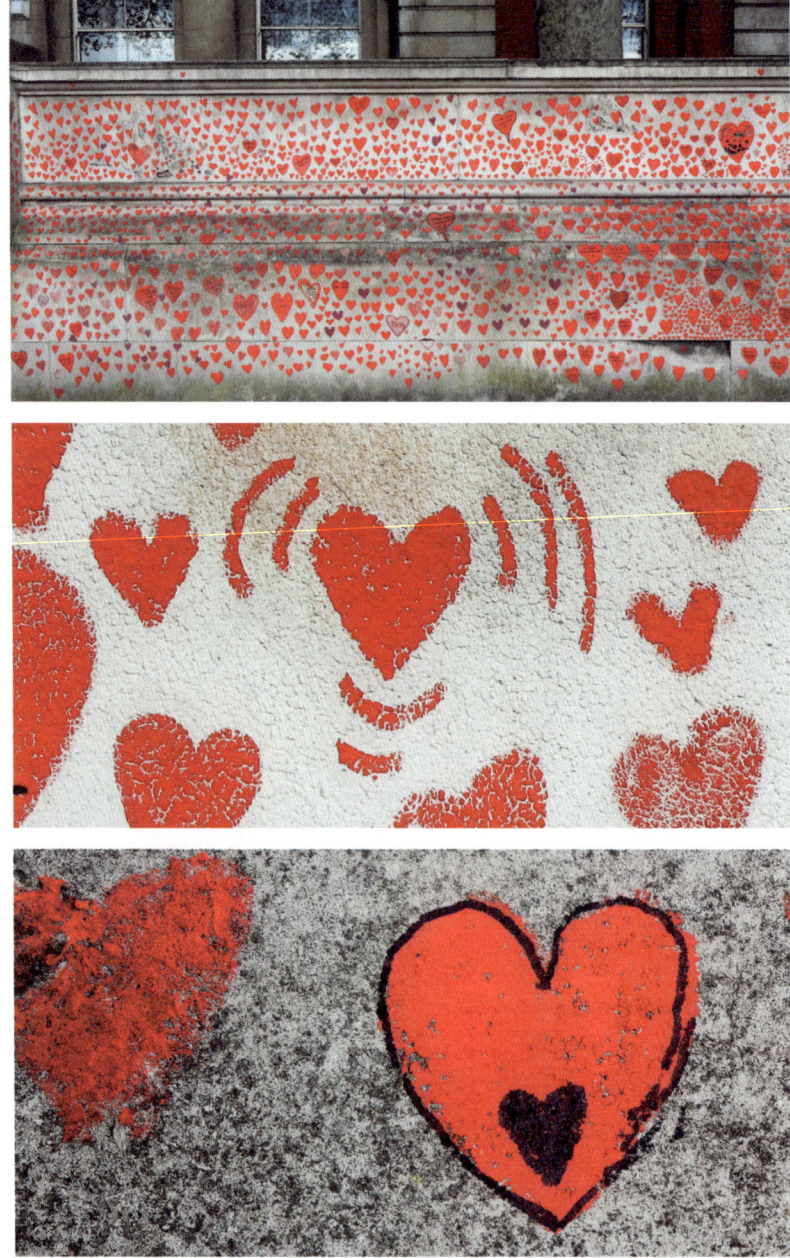

was perceived to have led to the fire and consequent loss of life. Under a flyover bridge for the A40 dual carriageway (Westway) that bisects that part of town, a large, spontaneous shrine has grown. The urban harshness of the concrete pylons and the carriageway they fix somehow form an apt contrasting backdrop for a very human memorial. Here, colourful images of hearts are found in abundance. They are used comprehensively and relentlessly in this moving commemorative setting. The diversity of heart styles is remarkable, but they are invariably unique interpretations of some familiar trope and, as such, they are accessible and understood.

Similarly, the London COVID-19 memorial in front of St Thomas' Hospital and running along the South Bank of the River Thames contains an enormous proliferation of heart emblems and motifs, many of which were added during the pandemic as an immediate and powerful commemoration of lost friends and relatives (Figure 9.3). The memorial is huge, extending over several hundred metres and featuring tens of thousands of heart images. There are beating hearts, breaking hearts, hearts within hearts, black spots in hearts, punctured hearts and bleeding hearts – a full repertoire. These are representations that we have already seen played out and recurring over centuries. It does not matter whether these particular heart motifs were learned and reinterpreted from previous exposure or whether they were intuited, in whole or in part, afresh. Whatever the precise origin of these images, the heart has been used powerfully, repeatedly, to denote themes that are consistent enough for us, as viewers, to have a clear idea of what was intended. The heart icon is in the vernacular.

In this final chapter, I want to examine how, and perhaps why, representations of the heart have prospered despite our understanding of its function in scientific terms. In other words, to touch on the heart's mystery. I would like to look at the origins of the heart motif

FIGURE 9.3 National COVID-19 Memorial Wall, South Bank of the River Thames, 2021, London.

and to question the elements that make a heart icon recognizable and meaningful. Is there a direct lineage from earlier representations or, perhaps, the possibility of reimagined attributes in a post-scientific age and context? While variously graphically depicted, dripping, punctured, compressed, flaming, fractured or throbbing, most of the contemporary heart images in this chapter are in the form of the 'heart motif'. Around the world, everyone knows what that is, but what are the origins of this familiar symmetrical, curved form with its pointed apex and scalloped, bilobed base (Figure 9.3)? In discussing the question of heart depiction, that is the question I have been asked by far the most often.

While the motif is now near-universally recognized, its precise origin has not been so easy to identify. In his book *The Shape of the Heart: A Contribution to the Iconology of the Heart*, Dutch cardiologist Pierre Vinken attributed the shape of the heart motif to a representation of the three-chambered heart of classical texts, in particular by Aristotle.[1] I am not convinced. We don't know why Aristotle believed the heart had three chambers (and not four); perhaps he mistook the right atrium for an extension of the caval veins to which it is attached. But, if anything, the heart motif is more suggestive of a symmetrical, two-chambered structure, rather than three. Jack Hartnell writes that the motif that we now associate with the heart was a common decorative element in the Middle Ages, and after 1450, with the availability of print, it became both more abundantly reproduced and associated with love.[2] Others have noted that the most likely 'inspiration' for the motif itself comes from nature. In his book *Christ to Coke: How Image Becomes Icon*, Professor Martin Kemp writes:

> It would be good to tell a tidy story of the origins and rise of the heart shape, but I am not convinced that it can be done. A number of claims have been made that the heart shape as a symbol of love originated with ancient plant symbolism. Stylised ivy leaves in various decorative contexts assumed heart shape to a greater or lesser degree.[3]

254

Certainly, the ivy leaf has been used as a symbol of enduring love, and one can find almost identical symbols all over the world, with small stems inscribed on Greek vases, Roman stonework and in early Bengali texts. It is not a huge jump from this shape to an imagined representation of the two upper chambers (atria) of the heart tapering towards the apex of the ventricles beneath. The image in Figure 2.6 shows an additional possibility, with the left and right sides of the heart separated by a groove or sulcus that in life couches the left coronary artery. The ventricles meet in a point at the apex, while the atria form the expanded bulbs at the base, neatly fusing the anatomical heart and the heart motif.

For a time, I was inclined towards the explanation tentatively offered by Kemp and others and had largely resigned myself to some obscure, anonymous, untraceable origin of the motif, though it would have been intellectually frustrating and perhaps even disappointing if the heart had made so little specific contribution to its own representation.

By the time I arrived in Egypt, primarily to study depictions of hearts on ancient papyrus scrolls in the *Book of the Dead* in the Cairo Museum, the manuscript for this book was largely finished. I had already collected hundreds of images of hearts, painted, pasted, daubed, scrawled or inscribed around the world – some of which appear here. My eye had become highly attuned to the heart motif form, trained by the habit of scanning my environment for interesting specimens. Then, on a column in the temple at Esna, on the River Nile, a few miles to the south of Luxor, I first saw a particular type of engraving that resembled the heart motif (Figure 9.5). The shape was immediately recognizable, but it stretched credulity to think that this was a representation of a heart itself, not least given the attached tubular structure rising from the top, which does not really have any convincing resemblance to the aorta. In fact, when scholars of Egyptian hieroglyphs first identified this symbol

FIGURE 9.4 (overleaf) Kalpasutra Manuscript, Victoria and Albert Museum, fifteenth to sixteenth centuries, India.

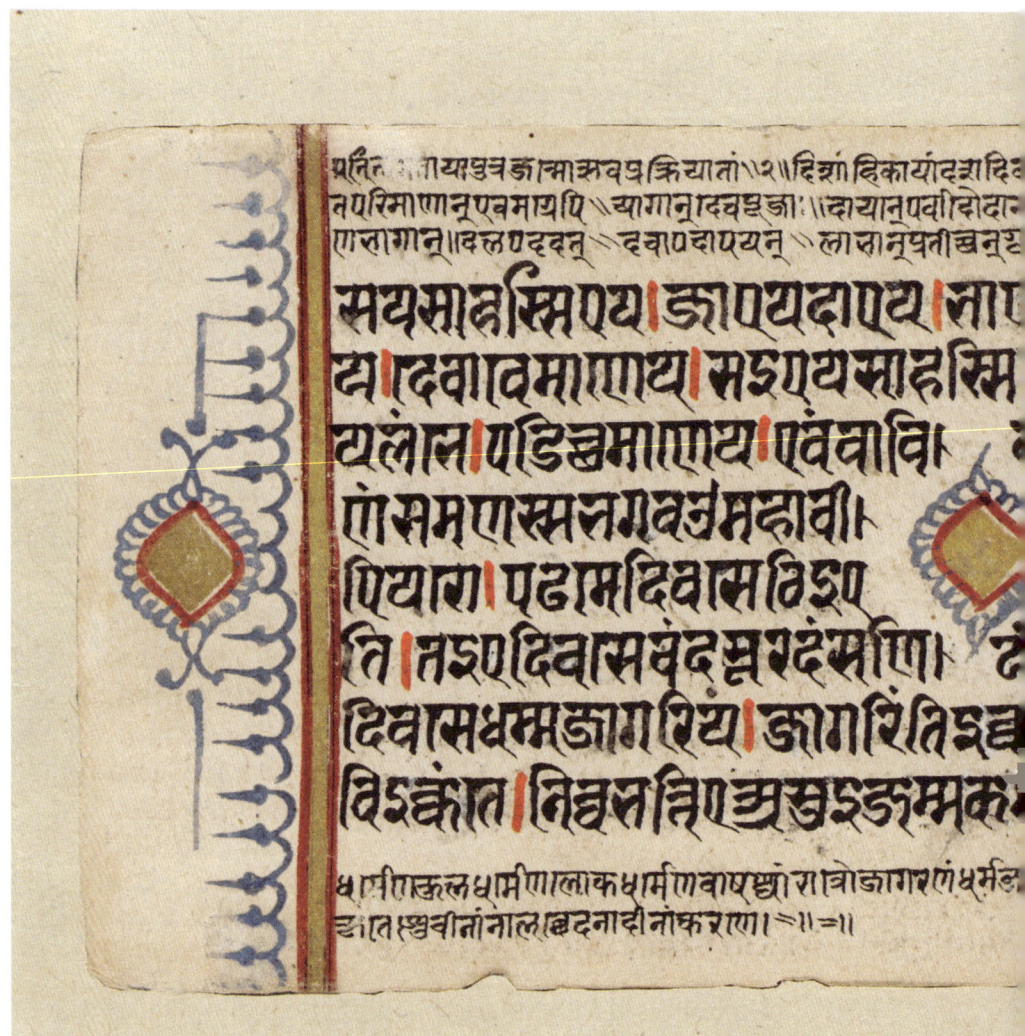

(*Nefer*, indicating 'beauty'), they thought that the symbol represented a lute, with its characteristic globular body and straight neck. However, latterly, it has become widely accepted that this is a representation of the heart. The tubular structure is not intended to represent a blood vessel but the trachea. In some versions the tube is decorated with striations, corresponding to the rings of cartilage on the trachea (which allow it flexibility but protect it from collapse, analogous to the ribbed exoskeleton in a vacuum cleaner hose). This is very plausibly the trachea, and it is made more plausible by the four branching side tubes that are inspired by the bifurcation of the left and right main bronchi. In reality, of course, the trachea shows no such connection to the heart. The heart is full of blood, not air. Gas exchange occurs in the lungs, where the thin membranes of the alveoli bring blood into close adjacency with the air to allow for the diffusion of oxygen and carbon dioxide – but the blood and air do not mix. However, the distorted anatomy fits well with the prevailing ancient Egyptian belief that the branching structures of the body, the *metu* (including the arteries and the veins), in life were filled not with blood but with air.[4] The co-representation of heart and aorta was not limited to ancient Egypt; we have already seen something strikingly similar in sixteenth-century European anatomy (Figure 4.2).

In the classification of hieroglyphs devised by the Egyptologist Sir Alan Gardiner and published in 1927, *Nefer* appears at F35 in the category 'parts of mammals'. So, in fact, the origins of the characteristic shape of the heart motif have been recognized for almost 100 years, but they have simply been confined to a particular field of study and have not spilled over into a wider appreciation. Of course, I cannot know for certain whether there is a direct, uninterrupted lineage from ancient Egypt to the contemporary heart motif, or how and even whether this motif might have trickled down through the ages – or for that matter why the Egyptians created this particular symbol to denote 'beauty'. It is at least possible that similar later representations were adopted or created independently, given the simplicity of the icon and its universality. That

FIGURE 9.5 The *Nefer* hieroglyph, Temple of Kom Ombo, Ptolemaic dynasty, 180–47 BCE, Egypt.

said, it seems unlikely that we will find earlier versions than this one, carved, quite literally, in ancient stone. On current evidence, I believe this to be the origin of the heart motif.

And so it is ironic – or perhaps not ironic at all, but pleasingly circular, given this putative origin – that the heart has reclaimed function as a hieroglyph, as in the phrase 'I ♥ NY' (Figure 9.6). It is difficult to bring to mind any symbol that can be similarly used (and recognized) when inserted as part of a written phrase. This logo was designed in 1977 as part of a campaign to promote tourism in New York state. Its creator, Milton Glaser, is said to have used a red crayon to sketch the design on the back of an envelope while riding in a taxi.[5] Almost immediately the logo gained widespread and instant recognition as a symbol of the city. In a comment in *The Village Voice* in 2011, Glaser remarked: 'I'm flabbergasted by what happened to this little simple nothing of an idea.'[6] But in a sense, even in his modesty, he may have missed the point. Inspired as the logo was, and undeniably successful, perhaps its success was not merely attributable to Glaser's neat, compact, upbeat design and crisp businesslike urban typeface, but also to his application of a universally embraced symbol that

FIGURE 9.6 Advertising logo for the State of New York, Milton Glaser, 1977, New York.

already suffused a global collective consciousness. His genius might not have been altogether in the design but more in the intuition.

There is one more small element that fascinates me in heart depictions: the black spot. This is a detail or embellishment added to heart representation and iconography, and we first came across it in the early anatomical images as the *nigrum granum* ('black grain') (Figures 2.2–2.4). The origin and purpose of the black spot in the anatomical images was not elaborated in the accompanying medieval text, but it was not drawn 'from life'; there is no anatomical correlate. What induced the illustrator to include the spot is unclear. Indeed, a similar inclusion in the heart is also found in Leonardo da Vinci's early anatomical drawings (Figure 3.1). Remarkably, even contemporary images of the heart, street art images from Paris, London and Rome, all show black dots or crosses in the heart. Again, there is no annotation or explanation of what the marks are intended to convey. Perhaps the artist did not need a literal translation; perhaps they assumed the meaning was contained within the image, or perhaps they did not care to explore the symbol with such deliberate analysis.

Among all the street art hearts I have stumbled upon, only once have I seen the artist in the process of making the work. A female artist – I guess in her twenties or thirties – waded through undergrowth on a disused railway line in north London. There, she found an expanse of unadulterated red brick cupped in a small, arched recess. As I watched, she began to spray first the outline of the heart in its anatomical form and then to add the blood vessels exiting the base of the heart, the coronary arteries that course along its surface and – to my astonishment on this ostensibly anatomical representation – an intense black spot. She completed the image with cartoon arcs to suggest a beating heart (Figure 9.7). Thrilled to have seen her make the piece, and having offered some compliments and explained my interest, I asked her why she had added the black spot. Somewhat to my surprise, she did not know – or at least, that is what she chose to tell me. These black spots on the heart

FIGURE 9.7 Street art heart being made, 2020, London.

or inclusions within the heart have been a particularly intriguing feature that crops up again and again over the centuries. Probably they do not stand for any one thing in particular; I imagine they symbolize, at least in their modern form, some stain, patch, hurt or wound within the perceived seat of feeling and emotion – in essence, a very simple concept. These are spontaneous expressions. In fact, even the quintessential heart motif was repurposed in this way. After the 9/11 atrocities at New York's World Trade Center, Glaser edited the original logo so that it read: 'I ♥ NY MORE THAN EVER', this time placing a dark black bruise on New York's heart.[7]

In the accompanying examples of street art, I have tried to select geographically diverse examples that nonetheless cluster together thematically. I have started to describe them and to attempt some superficial interpretation, but that would be completely missing the point on my part. If my reasoning is correct, the reader, certainly by this point, should need no signposting; the meanings are in the images themselves or, more precisely, the emoted response and any more considered interpretation are generated at the interface of the artist's rendering and the viewer's response, and each of these might in turn reflect some common experience of the heart.

But even as I contemplate the heart's lofty murmurings, I am conscious of the risk that its image could become so widespread and so overused that it loses meaning. Consider the range of heart emojis that offer themselves to our electronic scripts (Figure 9.8). Given all that I have said about the heart's very immediate association with emotion, about its use as an ancient and contemporary hieroglyph, there was an inevitability that it would be used in this new domain. In the 1980s, a trend began of using standard punctuation symbols to fashion representations of facial expressions. Smiling faces :-) and sad faces :-(were initially intended to identify text messages that were seriously intended or not. By the 1990s, Japanese mobile phones had moved on from these 'emoticons' and were starting to offer a new image-based lexicon of 'emojis'. The problem of translating emotions into dots and dashes and even tiny images is summed up here, succinctly and untainted by the merest pinch of irony:

> The fact that emojis are related to emotions and are becoming means of communication makes emoji prediction an interesting problem for Natural Language Processing (NLP). If we assume that an emoji is a label of the text corresponding to an emotion, then we would face the sentiment analysis problem. However, the classical sentiment analysis problem only analyses whether a text is positive or negative – sentiment polarities of sentences and documents.
>
> [...] Advanced models only have several additional emotions like happiness and anger. On the other hand, emoji classification has a larger population of candidates. As of November 2017 – there are 2623 emojis available in unicode. They are much more detailed and complicated to predict, because one emoji corresponds to many emotions based on its use and the same emotion can be expressed with various emojis. Sentiment analysis using emojis as emotional markers would be a tedious task.[8]

This menu of more than twenty heart emojis (Figure 9.8), though unlikely to be exhaustive, is still hard to fathom. Exactly what do all these symbols represent? Does this proliferation of hearts undermine my sense that representations of the heart are intimately linked to shared human experience? As expected, there are some black hearts and cracked hearts and punctured hearts and historic love hearts, but I struggle to intuit any meaning in a white heart in a purple box? What does a black heart rotated counterclockwise by ninety degrees or a pink heart decorated with golden sparkles insinuate? In fact, I suspect that we are seeing a selection of heart motifs that has not yet been subject to natural selection. I predict that the motifs that relate to the archetypal heart, built from shared experience, will prevail, while the others will fade to nothing over time. We might want to consider a distinction between the semantic and the hermeneutic in the contemporary usage of heart icons. In the semantic, the heart is an icon with a defined and accepted meaning. We have long been familiar with the 'I "love" New York' semiotic, but with the advent of the emoji, the possible interpretations have proliferated. The latter draws from a richer hermeneutic history of heart icons that this text has sought to explore.[9]

It might also be instructive to ask ourselves what features of the heart need to be captured in order to convey first that it is a 'heart' and second some characteristic of 'heartness'. In his seminal text *Art and Illusion*, Professor Ernst Gombrich tackles this type of question head on. In his example, he takes a circle and annotates it as a loaf of bread; adding a curve to the top transforms our perception to that of a bag, and with two further small triangular notches it resembles a purse; with the mere addition of a short arc at the bottom to denote a tail, the whole is unmistakably transformed into a cat seen from behind. These are recognizable stereotypic forms.[10]

In the very literal sense, we might look at any number of scrawled examples where, interpreted in context, even very primitive shapes might be recognized as heart shaped. In this vein, Willem de Kooning's *Valentine*

♥ [Alt] + [3]
♡ [Alt] + [9825]
❤ [Alt] + [10084]
🖤 [Alt] + [10085]
❣ [Alt] + [10083]
❦ [Alt] + [010 086]
❧ [Alt] + [010 087]
💑 U + 1F491
💓 U + 1F493
💔 U + 1F494
💕 U + 1F495
💖 U + 1F496
💗 U + 1F497
💘 U + 1F498
💙 U + 1F499
💚 U + 1F49a
💛 U + 1F49b
💜 U + 1F49c
💝 U + 1F49d
💞 U + 1F49e
💟 U + 1F49f
ღ [Alt] + [04326]

FIGURE 9.8 Heart emojis.

265

features a very loosely painted outline, with adjacent abstract forms, but that is nonetheless instantly recognizable as a heart (Figure 9.9). Figuratively crude, it still clearly denotes the heart motif. It is for the viewer to respond to the vaguely distorted black outline and its relationship to the other objects on the page. But ultimately, recognition relies on allusion to the recognized heart form.

On the other hand, Henri Matisse reduced the heart to an essentially unrecognizable form in his artwork *Icarus* (Figure 9.10). What could be less formed than a simple circle (only a line or a dot)? Nevertheless the position of the scarlet circle in Icarus' chest – carefully positioned and coloured against the backdrop of a bright blue monotone body and with knowledge of the context – clearly indicates a heart.

Completed in 1947, *Icarus* is one of twenty pieces Matisse included in his illustrated book *Jazz*. The image derives from a collage, which was then printed using the pochoir stencil technique. The viewer can determine the significance. Some imagine that: 'Matisse suggests that as a man rises toward the stars he does so not by means of his own powers, constrained by the mortal powers that define them, but rather from the power of the heart that beats in his chest',[11] while for others: 'For Matisse, this piece relates to the Greek myth of Icarus who "with a passionate heart, falls out of the starry sky".'[12]

To my eye, and from my heart, Icarus is falling as his heart thunders in terror. But the contradictions do not matter. The interest in Matisse's stripped-down heart is the mode of communication, requiring interpretation born from empathy. Matisse has reduced his representation of the heart to an extreme degree, shifting the interpretation of its significance heavily towards the viewer.

All this to say that the heart has become a visual icon. Precisely what that might mean is to a large degree subjective, open to interpretation. But I am persuaded by the definition offered by Professor Martin Kemp:

FIGURE 9.9 Willem de Kooning, *Valentine*, 1971, New York.

an iconic image is one that has achieved wholly exceptional levels of widespread recognizability and has come to carry a rich series of varied association for very large numbers of people across time and cultures, such that it has to a greater or lesser degree transgressed the parameters of its initial making, function, context and meaning.[13]

On the other hand, it's simply unnecessary to check off the heart's attributes as an icon against this list. As with so much in relation to the heart, we simply know it to be so.

FIGURE 9.10 Henri Matisse, *Icarus*, one of twenty plates from the book *Jazz*, 1947, Paris.

Acknowledgements

In this book, I have tried to tell a *story* of how the heart has been regarded across time and cultures and to draw out commonly held beliefs and attributes that have described that path. The project has its own story. Its earliest inception was in the senior common room of Balliol, where my friend and colleague Elena Lombardi, Professor of Italian literature, gave me a copy of Professor Heather Webb's spellbinding book, *The Medieval Heart*. I had pursued a career in cardiology based largely on my unexplored lifetime instinct that the heart was 'the most important bit'. Before reading Professor's Webb's book I had little sense of how deeply that notion was etched into Western cultural history.

Around the same time, Tim Wilson, Emeritus Professor of the Arts of the Renaissance, and then Keeper of Western Art at the Ashmolean Museum introduced me to Dr Jim Harris, Teaching Curator at the museum. Together, Jim and I organised a symposium *Seeing the Human Heart* (Ashmolean, 2013), which was generously supported by the British Heart Foundation and the National Institute for Health Research. After a long incubation, I had the good fortune to meet Georgina Capel and Anthony Cheetham, whose encouragement spurred me to develop a text. I'm grateful to Tom Baldwin and Rebecca Nicholson for bringing us together.

The shape of the book – and particularly the extension beyond my own often narrow viewpoints – has been fuelled by many rich conversations and travels with Jasmine Dellal and benefited from her unceasing curiosity, fascination for the non-obvious and long-held interest in traditions beyond the Western and mainstream. I'm grateful to her for many exchanges and challenges, sprinkled with her perceptive insights and perspectives. Her skills in narrative have helped to open and tell a story and to soften my harder, more mechanical and scientific, thinking and phrasing.

Listening to, and learning from, Professor Denis Noble over the years has been an education. The range and depth of this knowledge and his willingness to see outside, and beyond, conventional disciplines and modes of inquiry has been inspiring. I am extremely indebted to him for his sustained intellectual inspiration, guidance and generosity. Countless contributions have enriched the text. I am particularly indebted to Dr Daryl Green (for sharing images as part of the "*Seeing in the 3-D exhibition*") and to Sandy Rower and the Calder Foundation for generously delving into their archive. I owe gratitude to Arlene Shaner, librarian at the New York Academy of Medicine who sourced and laid out a stream of exquisite and relevant historic texts. Thanks also to Mark Statham at Caius College library for helping me to view the early English anatomy manuscripts (Gonville and Caius 190/223) and to the staff of the library of the Royal Society of Medicine, the Bodleian Library and of the Wellcome Collection.

I am grateful to Professor Philip Kilner, Professor Michael Markl, Dr Patrick Hales, Dr Malenka Bissell and Dr Erica Dall'Armellina for contemporary MR images. Many have fed me information, nuggets, leads, texts, links, images and wise words. I am grateful to Lucy Atkins, Eunice Berry, Nick Laird, Dr Taylor McCall, Professor Simon May, Latha Menon, Christopher Quilkey, Elizabeth Sheinkman, Professor Diego Zancani, Adam Zuabi. The translations by Dr Lorenzo Caravaggio (from Latin) and Sophia Choudhury (Spanish) were invaluable and are included. I am very grateful for the expert comments, gentle critiques, observations and edits on the pre-final versions of the manuscript from Professor Elena Lombardi, Professor Simon May, Dr Eric Southworth, Dr Jack Hartnell, Jasmine Dellal, Dr Bert Choudhury and Dr Nick Haining. While they have rounded off some of my coarser blunders, any errors that remain are mine.

Richard Milbank, Editor at Head of Zeus has been a pleasure to work with. The text has been vastly improved by his insightful and sympathetic working and sculpting, for which my very sincere thanks.

Also, to Clémence Jacquinet, Ellie Jardine, Simon Michele (cover) and Kate Wands, at Head of Zeus.

Finally, I would like to thank my family. I acknowledge with deep gratitude the enduring influence of my parents the late Sue Choudhury and the late Jack Choudhury FRCS who, among many other things, valued breadth and diversity in thought and learning, brought books to our lives and gave me my first texts on art. In the comments, above, I have noted my appreciation for all that Jasmine Dellal has brought to this text. I am profoundly grateful to her for so much more. Thank you, Bert, Maya, Sophia and Milo, for your inspiration and support – I love and appreciate you and delight in our shared quest to be K, C and C.

Notes

Chapter 1

1. Leroi, A. M. *The Lagoon: How Aristotle Invented Science* (New York: Viking, 2014).
2. Taylor, J. H. *Journey Through the Afterlife: Ancient Egyptian Book of the Dead* (London: British Museum Press, 2011).
3. Andrews, C. *Amulets of Ancient Egypt* (Austin: University of Texas Press, 1994).
4. Young, L. 'The Human Heart, An Overview', in J. Peto (ed.), *The Heart* (Newhaven: Yale University Press, 2007).
5. Charbonnneau-Lassay, L. 'The Human Heart and the Notion of the Heart of God in the Religion of Ancient Egypt', *Regnabit* (1925).
6. Pierozak, G. 'Regnabit. Universal Review of the Sacred Heart', Oriens, 84–6 (2011).
7. Kerkhove, R. 'Dark Religion? Aztec Perspectives on Human Sacrifice', *Sydney Studies in Religion* (2008).
8. Tiesler, V., Cucina, A. 'Procedures in Human Heart Extraction and Ritual Meaning', *Latin American Antiquity* 17 (2006).
9. Olivelle, P. 'Heart in the Upanishads', *Rivista di Studi Sudasiatici* I (2006).
10. Radhakkrishnan, S. *The Principal Upanishads* (New Delhi: HarperCollins, 1994).
11. Guénon, R. 'The Symbolism of the Heart', in *Fundamental Symbols: The Universal Language of Sacred Science* (London: Quinta Essentia, 1962).
12. Ibid.
13. Ibid.
14. Rodriguez Munoz, B. *Ayurvedic Man* (London: The Wellcome Collection, 2017).
15. Guénon, R. 'The Symbolism of the Heart', in *Fundamental Symbols: The Universal Language of Sacred Science* (London: Quinta Essentia, 1962).
16. Rodriguez Munoz, B. *Ayurvedic Man* (London: The Wellcome Collection, 2017).
17. Ibid.
18. Ibid.
19. Jinsheng, Z. 'Observational Drawing and Fine Art in Chinese Materia Medica Illustration', in V. Lo, P. Barrett P (eds), *Imagining Chinese Medicine* (Leiden: Brill, 2018).
20. Shumin, W., Fuentes, G. 'Chinese Medical Illustration: Chronologies and Categories', in V. Lo, P. Barrett (eds), *Imagining Chinese Medicine* (Leiden: Brill, 2018).
21. Ibid.
22. Bivins, R. 'Imagining Acupuncture: Images and the Early Westernisation of Asian Medical Expertise', in V. Lo, P. Barrett (eds), *Imagining Chinese Medicine* (Leiden: Brill, 2018).
23. Choulant, L. *Chinese Anatomy: History and Bibliography of Anatomic Illustration* (Leipzig: Rudolph Weigel, 1852).
24. Ibid.
25. Appiah, K. A. 'Why Africa? Why Art?', in T. Philips (ed.), *Africa: The Art of a Continent* (London: Royal Academy of Arts, 1995).

26. Odorisio, D. M. 'The Alchemical Heart: A Jungian Approach to the Heart Center in the Upanishads and in Eastern Christian Prayer', *International Journal of Transpersonal Studies* 33 (2014).

27. Ibid.

Chapter 2

1. McCall, T. 'Reliquam dicit pictura: Text and Image in a Twelfth-Century Illustrated Anatomical Manual (Gonville and Caius College, Cambridge, MS 190/223)', *Transactions of the Cambridge Bibliographical Society* xvi (2016).

2. Osler, W. *The Evolution of Modern Medicine* (New Haven: Yale University Press, 1921).

3. McCall, T. 'Reliquam dicit pictura: Text and Image in a Twelfth-Century Illustrated Anatomical Manual (Gonville and Caius College, Cambridge, MS 190/223)', *Transactions of the Cambridge Bibliographical Society* xvi (2016).

4. Ibid.

5. Ibid.

6. Hill, B. H. *The Fünfbilderserie and Medieval Anatomy* (University of North Carolina at Chapel Hill, 1963).

7. McCall, T. Personal communication. 2019.

8. Savage-Smith, E. *Mansur ibn Ilyas. Tashrih-i badan-i insan* [*Anatomy of the Human Body*], Iran, *c.* 1390 (2016).

9. Guénon, R. 'The Symbolism of the Heart', in *Fundamental Symbols: The Universal Language of Sacred Science* (London: Quinta Essentia, 1962).

10. Jager, E. *The Book of the Heart* (Chicago: The University of Chicago Press, 2000).

11. Ibid.

12. Ibid.

13. Ibid.

14. Park, K. 'Autopsy and Dissection', in J. J. Martin (ed.), *The Renaissance Italy and Abroad Rewriting Histories* (London: Routledge, 2003).

15. Hartnell, J. *Medieval Bodies* (London: The Wellcome Collection, 2018).

16. Webb, H. *The Medieval Heart* (Yale University Press, New Haven and London, 2010).

17. Ibid.

18. Hartnell, J. *Medieval Bodies* (London: The Wellcome Collection, 2018).

19. May, S. *Love: A History* (London: Yale University Press, 2011).

20. Hartnell, J. *Medieval Bodies* (London: The Wellcome Collection, 2018).

21. Jones, J. *Balliol College: A History 1263–1939* (Oxford: Oxford University Press, 1988).

22. McInerny, R., O'Callaghan, J. *Saint Thomas Aquinas.* E. N. Zalta (ed.) (Stanford, USA: Metaphysics Research Lab, Stanford University, 2018).

23. Larkin, V. R. 'St. Thomas Aquinas on the Movement of the Heart', *Journal of the History of Medicine and Allied Sciences* 15 (1960).

24. Aquinas, T. *De Motu Cordis* (1270), available from: https://isidore.co/aquinas/DeMotuCordis.htm

Chapter 3

1. Hunter, W. *Introductory Lectures Delivered by Dr William Hunter to His Last Course of Anatomical Lectures*

at his *Theatre in Windmill Street*
(London: Printed by Order of the
Trustees of J Johnson, 1784).

2. Freud, S. *Leonardo da Vinci. A
   Psychosexual Study of an Infantile
   Reminiscence* (New York: Moffat,
   Yard and Co., 1916).

3. Gombrich, E. H. *In Search of
   Cultural History, Ideals and Idols*
   (Oxford: Phaidon, 1979).

4. Vasari, G. *Lives of the Artists* (New
   York, Harry N Abrams, 1979).

5. Clayton, M., Philo, R. *Leonardo da
   Vinci: Anatomist* (London: Royal
   Collection Trust, 2012).

6. Keele, K. *Leonardo da Vinci and the
   Art of Science* (Hove, England: Priory
   Press, 1977).

7. Clayton, M., Philo, R. *Leonardo da
   Vinci: Anatomist* (London: Royal
   Collection Trust, 2012).

8. Keele, K., Pedretti, C. *Leonardo
   da Vinci: Corpus of the Anatomical
   Studies in the Collection of Her
   Majesty the Queen at Windsor
   Castle* (London: Johnson Reprint
   Company, 1980).

9. Clayton, M., Philo, R. *Leonardo da
   Vinci: Anatomist* (London: Royal
   Collection Trust, 2012).

10. Hunter, W. *Introductory Lectures
    Delivered by Dr William Hunter to
    His Last Course of Anatomical Lectures
    at his Theatre in Windmill Street*
    (London: Printed by Order of the
    Trustees of J Johnson, 1784).

11. Keele, K., Pedretti, C. *Leonardo
    da Vinci: Corpus of the Anatomical
    Studies in the Collection of Her
    Majesty the Queen at Windsor
    Castle* (London: Johnson Reprint
    Company, 1980).

12. Ibid.

13. Clayton, M. *Leonardo da Vinci: One
    Hundred Drawings from the Collection
    of Her Majesty the Queen* (London:
    Merrell Holberton, 1996).

14. Keele, K., Pedretti, C. *Leonardo
    da Vinci: Corpus of the Anatomical
    Studies in the Collection of Her
    Majesty the Queen at Windsor
    Castle* (London: Johnson Reprint
    Company, 1980).

15. Ibid.

16. Kemp, M. *Leonardo da Vinci: The
    Marvellous Works of Nature and Man*
    (Oxford: Oxford University Press,
    2006).

17. Clayton, M., Philo, R. *Leonardo da
    Vinci: Anatomist* (London: Royal
    Collection Trust, 2012).

18. Keele, K. *Leonardo da Vinci:
    Anatomical Drawing from the Royal
    Collection* (London: Royal Academy
    of Arts, 1977).

19. Noble, D., DiFrancesco, D.,
    Zancani, D. 'Leonardo da Vinci
    and the Origin of Semen', *Notes and
    Records of the Royal Society* 68 (2014).

20. Clayton, M., Philo, R. *Leonardo da
    Vinci: Anatomist* (London: Royal
    Collection Trust, 2012).

21. Keele, K., Pedretti, C. *Leonardo
    da Vinci: Corpus of the Anatomical
    Studies in the Collection of Her
    Majesty the Queen at Windsor
    Castle* (London: Johnson Reprint
    Company, 1980).

22. Clayton, M. *Leonardo da Vinci: One
    Hundred Drawings from the Collection
    of Her Majesty the Queen* (London:
    Merrell Holberton, 1996).

23. Bellhouse, B. J., Bellhouse, F. H.
    'Mechanism of Closure of the Aortic
    Valve', *Nature* 217(5123) (1968).

24. Bissell, M. M., Dall'Armellina, E.,
    Choudhury, R. P. 'Flow Vortices in
    the Aortic Root: In Vivo 4D-MRI

Confirms Predictions of Leonardo da Vinci', *European Heart Journal* 35(20) (2014).

Chapter 4

1. Harvey, W. *Exercitatio Anatomica de Motu Cordis et Sanguinis in Animalibus* (London, 1628).
2. Jardine, L. 'The New Organon: Introduction', in L. Jardine, M. Silverthorne (eds), *The New Organon Francis Bacon* (Cambridge: Cambridge University Press, 2000).
3. Huxley, T. *Fortnightly Review* (London, February 1878).
4. Park, K. 'Autopsy and Dissection', in J. J. Martin (ed.), *The Renaissance Italy and Abroad Rewriting Histories* (London: Routledge, 2003).
5. Laurenza, D. *Art and Anatomy in Renaissance Italy* (New York: The Metropolitan Museum of Art, distributed by Yale Press, 2012).
6. Choulant, L. *History and Bibliography of Anatomic Illustration* (Leipzig: Rudolph Wiegel, 1852).
7. Pantin, I. 'Analogy and Difference: A Comparative Study of Medical And Astronomical Images in Books 1470-1550', in N. Jardine, I. Fay (eds), *Observing the World through Images Diagrams and Figures in the Early Modern Arts and Sciences* (Leiden: Brill, 2014).
8. Wear, A. 'Early Modern Europe 1500-1700', in L. I. Conrad, M. Neve, V. Nutton, R. Porter, A. Wear (eds), *The Western Medical Tradition 800 BC to 1800 AD* (New York: Cambridge University Press, 1995).
9. Osler, W. *The Evolution of Modern Medicine* (New Haven: Yale University Press, 1921).
10. Lander, E. S., Linton, L. M., Birren, B., Nusbaum, C., Zody, M. C., Baldwin, J., et al. 'Initial Sequencing and Analysis of the Human Genome', *Nature* 409(6822) (2001).
11. Malloch, A. *William Harvey* (New York: Paul B. Hoeber, 1929).
12. Ibid.
13. Saunders, J. B. dC. M., O'Malley, C. D. *The Illustrations from the works of Andreas Vesalius of Brussels* (Cleveland and New York: The World Publishing Company, 1950).
14. Malloch, A. *William Harvey* (New York: Paul B. Hoeber, 1929).
15. Gregory, A. *Harvey's Heart: The Discovery of the Circulation* (Reading, UK: Cox and Wyman Ltd, 2001).
16. Vesalius, A. *De Humani Corporis Fabrica Libri Septum*. W. F. Richardson (trans.) (Novato, California: Norman Publishing, 1553).
17. Gregory, A. *Harvey's Heart: The Discovery of the Circulation* (Reading, UK: Cox and Wyman Ltd, 2001).
18. Choulant, L. *History and Bibliography of Anatomic Illustration* (Leipzig: Rudolph Wiegel, 1852).
19. Osler, W. *The Evolution of Modern Medicine* (New Haven: Yale University Press, 1921).
20. Ibid.
21. Saunders, J. B. dC. M., O'Malley, C. D. *The Illustrations from the works of Andreas Vesalius of Brussels* (Cleveland and New York: The World Publishing Company, 1950).
22. Ibid.
23. Carman, J. *Preface: On the Fabric of the Human Body* (Novato, California: Norman Publishing, 2009).
24. Gregory, A. *Harvey's Heart: The Discovery of the Circulation* (Reading,

UK: Cox and Wyman Ltd, 2001).

25. Wright, T. *Circulation: William Harvey's Revolutionary Idea* (London: Vintage, 2013).

26. Harvey, W. *Exercitatio Anatomica de Motu Cordis et Sanguinis in Animalibus* (London, 1628).

27. Ibid.

28. Malloch, A. *William Harvey* (New York: Paul B. Hoeber, 1929).

29. Wear, A. 'Early Modern Europe 1500-1700', in L. I. Conrad, M. Neve, V. Nutton, R. Porter, A. Wear (eds), *The Western Medical Tradition 800 BC to 1800 AD* (New York: Cambridge University Press, 1995).

30. Aubrey, J. *Brief Lives* (London: Folio Society, 1975, 1675).

31. Ekholm, E. 'Anatomy, Bloodletting and Emblems: Interpreting the Title-Page of Nathaniel Highmore's *Disquitio* (1651)', in N. Jardine, I. Fay (eds), *Observing the World through Images: Diagrams and Figures in the Early-Modern Arts and Sciences* (Leiden: Brill, 2014).

32. Ibid.

33. Voigts, L. E., McVaugh, M. R. 'A Latin Technical Phlebotomy and Its Middle English Translation', *Transactions of the American Philosophical Society* 74 (1984).

34. Descartes R. *Discours de la méthode pour bien conduire sa raison, & chercher la vérité dans les sciences. Plus La Dioptrique. Les Meteores. Et La Geometrie. Qui sont des essais de cete Methode* (Leiden, 1637).

35. Heitsch, D. 'Descartes, Cardiac Heat, and Alchemy', *Ambix* 63(4) (2016).

36. Ibid.

37. Heitsch, D. 'Descartes, Cardiac Heat, and Alchemy', *Ambix* 63(4) (2016).

38. Fye, W. B. 'Profiles in Cardiology: Rene Descartes', *Clinical Cardiology* 26(1) (2003).

39. Anstey, P. 'Descartes' Cardiology and Its Reception in English Physiology', in S. Gaukroger, J. Schuster, J. Sutton (eds), *Descartes' Natural Philosophy* (New York: Routledge, 2000).

Chapter 5

1. Malloch, A. *William Harvey* (New York: Paul B. Hoeber, 1929).

2. Jardine, L. *On a Grander Scale: The Outstanding Career of Sir Christopher Wren* (London: HarperCollins, 2002).

3. Aubrey, J. *Brief Lives* (London: Folio Society, 1975, 1675).

4. Frank, R. G. *Medicine: The History of the University of Oxford Volume IV Seventeenth Century Oxford*, IV (Oxford: Clarendon Press, 1997).

5. Hughes, J. T. *Thomas Willis 1621–1675: His Life and Work*, 2nd edition (Oxford: Rimes House, 2009).

6. Ibid.

7. Dewhurst, K. *Thomas Willis's Oxford Lectures* (Yarnton: Sandford Publications, 1980).

8. Molnar, Z. 'Thomas Willis (1621–1675), the Founder of Clinical Neuroscience', *Nature Reviews Neuroscience* 5(4) (2004).

9. Frank, R. G. *Medicine: The History of the University of Oxford Volume IV Seventeenth Century Oxford*, IV (Oxford: Clarendon Press, 1997).

10. Dewhurst, K. *Thomas Willis's Oxford Lectures* (Yarnton: Sandford Publications, 1980).

11. Ibid.

12. Frank, R. G. *Medicine: The History of the University of Oxford Volume*

IV Seventeenth Century Oxford, IV (Oxford: Clarendon Press, 1997).

13. Dewhurst, K. *Thomas Willis's Oxford Lectures* (Yarnton: Sandford Publications, 1980).

14. Zimmer, C. *Soul Made Flesh: The Discovery of the Brain and How It Changed the World* (London: Heinemann, 2004).

15. O'Connor, J. P. B. 'Thomas Willis and the Background to *Cerebri Anatome*', *Journal of the Royal Society of Medicine* 96 (2003).

16. Dewhurst, K. *Thomas Willis's Oxford Lectures* (Yarnton: Sandford Publications, 1980).

17. Sennett, R. *The Craftsman* (London: Allen Lane, 2008).

18. *The Concise Dictionary of National Biography* (Oxford: Oxford University Press, 1992).

19. Lower, R. *Tractatus de Corde* (London, 1669).

20. Ibid.

21. The Wellcome Collection, *Heartstrings*, 2013. Available from: wellcomecollection.org/works/rxt3cx88; accessed 7 Jan. 2024.

22. Kilner, P. J., Yang, G. Z., Wilkes, A. J., Mohiaddin, R. H., Firmin, D. N., Yacoub, M. H. 'Asymmetric Redirection of Flow through the Heart', *Nature* 404(6779) (2000).

23. Fye, W. B. 'Profiles in Cardiology: Rene Descartes', *Clinical Cardiology* 26(1) (2003).

24. Anstey, P. 'Descartes' Cardiology and Its Reception in English Physiology', in S. Gaukroger, J. Schuster, J. Sutton (eds), *Descartes' Natural Philosophy* (New York: Routledge, 2000).

25. Lower, R. *Tractatus de Corde* (London, 1669).

26. Donaldson, I. M. 'Cerebri Anatome: Thomas Willis and His Circle', *Journal of the Royal College of Physicians of Edinburgh* 40(3) (2010).

27. Pearce, J. M. 'Malpighi and the Discovery of Capillaries', *European Neurology* 58(4) (2007).

Chapter 6

1. Fye, W. B. 'Antonio Scarpa', *Clinical Cardiology* 20(4) (1997).

2. Ibid.

3. Jones, J. *Balliol College: A History 1263–1939* (Oxford: Oxford University Press, 1988).

4. Fye, W. B. 'Antonio Scarpa', *Clinical Cardiology* 20(4) (1997).

5. Gunther, R. T. *Early Science in Oxford: IX De Corde by Richard Lower, London 1669* (Oxford: The University Press, 1932).

6. Kilroy-Ewbank, L. G. 'Holy Organ or Unholy Idol? Forming a History of the Sacred Heart in New Spain', *Colonial Latin American Review* 23 (2014).

7. Morgan, D. *The Sacred Heart of Jesus: The Visual Evolution of a Devotion* (Amsterdam: Amsterdam University Press, 2008).

8. Kilroy-Ewbank, L. G. 'Holy Organ or Unholy Idol? Forming a History of the Sacred Heart in New Spain', *Colonial Latin American Review* 23 (2014).

9. Smith, J. H. *Joseph de Gallifet: The Catholic Encyclopaedia* (New York: Robert Appleton, 1909).

10. Kilroy-Ewbank, L. G. 'Holy Organ or Unholy Idol? Forming a History of the Sacred Heart in New Spain', *Colonial Latin American Review* 23 (2014).

11. de Gallifet, J. *De cultu sacrosancti cordis Dei ac Domini Nostri Jesu Christi* (Rome 1726).

12. Kilroy-Ewbank, L. G. 'Holy Organ or Unholy Idol? Forming a History of the Sacred Heart in New Spain', *Colonial Latin American Review* 23 (2014).

13. Barthes, R. *Sade, Fourier, Loyola* (London: Cape, 1977).

14. Marin, S. 'The Fleshy Heart of Jesus', *California Italian Studies*, 2015.

15. Feiler, T. 'Sacred Hearts and Pumps: Cardiology and the Conflicted Body Politic (1500–1900)', *Medical Humanities* 46(4) (2020).

16. Ibid.

17. Kilroy-Ewbank, L. G. 'Holy Organ or Unholy Idol? Forming a History of the Sacred Heart in New Spain', Brill; 2018.

18. Fraser-Giffords, G. *The Iconography of Mexican Folk Retablos* (University of Arizona, 1969).

19. Ibid.

20. Montenegro, R. *Mexican Painting (1800–1860)* (New York and London: Appleton-Century Company, 1933).

21. Kelman, P. *Baroque and Rococo in Latin America* (New York: Macmillan, 1951).

22. Brenner, A. *Idols behind Altars* (New York: Payson and Clarke Ltd, 1929).

Chapter 7

1. Herrera, H. *Frida: The Biography of Frida Kahlo* (London: Bloomsbury, 2003).

2. Ibid.

3. Ibid.

4. Ibid.

5. Westheim, P. 'Frida Kahlo: Una investigación estética', *Novedades: México en la Cultura 1952* (10 June 1951).

6. Herrera, H. 'Frida Kahlo Returns', in D. Sileno (ed.), *Frida Kahlo: Beyond the Myth* (Milan: 24 Ore Cultura, 2018).

7. Alfano Miglietti, F. 'The Mutating Identities of Frida Kahlo', in D. Sileo (ed.), *Frida Kahlo: Beyond the Myth* (Milan: 24 Ore Cultura, 2018).

8. Mahon, A. 'The Lost Secret: Frida Kahlo and the Surrealist Imaginary', *Journal of Surrealism and the Americas* 5 (2011).

9. Herrera, H. 'Frida Kahlo Returns', in D. Sileno (ed.), *Frida Kahlo: Beyond the Myth* (Milan: Mudec, 2018).

10. Ibid.

11. De Lee, J. B. *The Principles and Practice of Obstetrics* (Philadelphia, London: W. B. Saunders Co., 1913).

12. Westheim, P. 'Frida Kahlo: Una investigación estética', *Novedades: México en la Cultura 1952* (10 June 1951).

13. Herrera, H. 'Frida Kahlo Returns', in D. Sileno (ed.), *Frida Kahlo: Beyond the Myth* (Milan: Mudec, 2018).

14. Sileo, D. *Frida Kahlo: Beyond the Myth* (Milan: 24 Ore Cultura, 2018).

15. Kozloff., J. 'Frida Kahlo', *Women's Studies* 6 (1978).

16. Tiesler, V., Cucina, A. 'Procedures in Human Heart Extraction and Ritual Meaning: A Taphonomic Assessment of Anthropogenic Marks in Classic Maya Skeletons', *Latin American Antiquity* 17 (2006).

17. Clarke, J. A. *Becoming Edvard Munch: Influence, Anxiety and Myth* (New Haven: Yale University Press, 2009).

18. Prideaux, S. *Edvard Munch: Behind the Scream* (New Haven: Yale

University Press, 2005).

19. Morehead, A. 'Madness as Method: The Pathological Experiments of Edvard Munch', in *Nature's Experiments and the Search for Symbolist Form* (Pennsylvania: The Pennsylvania State University, 2017).

20. Morehead, A. *Symbolism and Nature's Experiments: Nature's Experiments and the Search for Symbolist Form* (Pennsylvania: The Pennsylvania State University Press, 2017).

21. Dall'Armellina, E., Karamitsos, T. D., Neubauer, S., Choudhury, R. P. 'CMR for characterization of the myocardium in acute coronary syndromes', *Nature Reviews Cardiology* 7(11) (2010).

22. Brandtzaeg, K. *Tracey Emin Meets Edvard Munch. The Loneliness of the Soul* (London: Royal Academy, 2021).

23. Hermann, M. D. 'Love Is Love', in *Andy Warhol Love Sex and Desire Drawings 1950 to 1962* (Cologne: Taschen, 2020).

24. Ibid.

25. Gopnik, B. 'Warhol's Defiant Hopes for Queer Art', in *Andy Warhol Love Sex and Desire Drawings 1950 to 1962* (Cologne: Taschen, 2020).

26. Gopnik, B. *1955–1956. Warhol* (New York: HarperCollins, 2020).

27. S. P., 'About Art and Artists; Five One-Man Shows Display Work by Painters from France, Spain and US', *New York Times*, 1956. Available from: timesmachine.nytimes.com/timesmachine/1956/03/03/86537818.html; accessed 7 Jan. 2024.

28. Calder, A. *Comment réaliser l'art? Abstraction-Création, Art Non Figuratif* (1932).

29. Perl, J. *Calder: The Conquest of Time: The Early Years: 1898–1940 (A Life of Calder)* (Newhaven: Yale University Press, 2017).

30. Rose, B. *Calder after the War* (London: PACE, 2013).

31. Perl, J. *Calder: The Conquest of Time: The Early Years: 1898–1940 (A Life of Calder)* (Newhaven: Yale University Press, 2017).

32. Sartre, J. P. *Alexander Calder: Les Mobiles de Calder*. Exhibition Catalogue*: Mobiles, Stabiles, Constellations* (Galerie Louis Carré, 1946).

33. Curtis. P. *Performance or Post-Performance: Alexander Calder Performing Sculpture* (London: Tate Publishing, 2015).

34. Sartre, J. P. *Alexander Calder: Les Mobiles de Calder*. Exhibition Catalogue*: Mobiles, Stabiles, Constellations* (Galerie Louis Carré, 1946).

35. Miller, A. Alexander Calder. Memorial Service. Whitney Museum of American Art, in Art. WMoA, editor. New York, 1976.

Chapter 8

1. Manders, K. 'The Magic of Numbers and Motion: The Scientific Career of René Descartes', *Philosophy of Science* 62 (1995).

2. Lower, R. *Tractatus de Corde* (London, 1669).

3. Coraboeuf, E., Weidmann, S. 'Potential de repos et potentiels d'action du muscle cardiaque, mesurés á l'aide d'électrodes intracellulaires', *Comptes rendus des séances de la Société de biologie et de ses filiales et associées*, 143 (1949).

4. Kleber, A. G., Niggli, E., McGuigan, J.A., Weingart, R. 'The Early Years of

Cellular Cardiac Electrophysiology and Silvio Weidmann (1921–2005)', *Heart Rhythm* 3(3) (2006).

5. Noble, D. *The Music of Life* (Oxford: Oxford University Press, 2006).

6. Hutter, O. F., Trautwein, W. 'Effect of Vagal Stimulation on the Sinus Venosus of the Frog's Heart', *Nature* 176(4480) (1955).

7. Noble, D. *The Music of Life* (Oxford: Oxford University Press, 2006).

8. Noble, D. 'Cardiac Action and Pacemaker Potentials Based on the Hodgkin-Huxley Equations', *Nature* 188 (1960).

9. Watson, J. D. *The Double Helix, A Personal Account of the Discovery of the Structure of DNA* (London: Weidenfeld and Nicolson, 1968).

10. Oakes, P. C., Fisahn, C., Iwanaga, J., DiLorenzo, D., Oskouian, R. J., Tubbs, R. S. 'A History of the Autonomic Nervous System: Part I: From Galen to Bichat', Child's Nervous System 32(12) (2016).

11. Salamone, P. 'Dynamic Neurocognitive Changes in Interoception after Heart Transplant', *Brain Communications* 2 (2020).

12. Templin, C., Ghadri, J. R., Diekmann, J., Napp, L. C., Bataiosu, D. R., Jaguszewski, M., et al. 'Clinical Features and Outcomes of Takotsubo (Stress) Cardiomyopathy', *New England Journal of Medicine* 373(10) (2015).

Chapter 9

1. Vinken, P. *The Shape of the Heart: A Contribution to the Iconology of the Heart* (Amsterdam: Elsevier, 1999).

2. Hartnell, J. *Medieval Bodies* (London: The Wellcome Collection, 2018).

3. Kemp, M. *Christ to Coke: How Image Becomes Icon* (Oxford: Oxford University Press, 2011).

4. 'Heart and Arteries', in Strouhal E., Vachala, B., Vymazalová H. (ed.), *The Medicine of the Ancient Egyptians* (Cairo: American University in Cairo Press, 2021).

5. Grimes, W. 'Milton Glaser, Master Designer of 'I ♥ NY' Logo, Is Dead at 91', *New York Times* (2020).

6. Doll, J. 'Milton Glaser on New Yorkers: "For Better or Worse You're Here, and Doomed to Be Here"', *The Village Voice* (November 23, 2011).

7. Grimes, W. 'Milton Glaser, Master Designer of 'I ♥ NY' Logo, Is Dead at 91', *New York Times* (2020).

8. Doliashvili, M., Ogawa, M.-C. B., Crosby, M. E. (eds), *Understanding Challenges Presented Using Emojis as a Form of Augmented Communication* (Copenhagen, Augmented Cognition Theoretical and Technological Approaches, 2020).

9. Klein, R. *Thoughts on Iconography, Form and Meaning* (Princeton, New Jersey: Princeton University Press, 1979).

10. Gombrich, E. H. *Art and Illusion. A Study in the Psychology of Pictorial Representation* (Washington: Pantheon, 1960).

11. Bordin, G., Polo D'Ambrosio, L. *Medicine in Art (A Guide to Imagery)* (Los Angeles: J. Paul Getty Museum, 2010).

12. Anonymous. *Icarus*, 2020. Available from: www.matissepaintings.org/icarus/; accessed 7 Jan. 2024.

13. Kemp, M. *Christ to Coke: How Image Becomes Icon* (Oxford: Oxford University Press, 2011).

Image Credits

p. 7. [top] © Fine Art Image/Heritage Image Partnership Ltd. / Alamy Stock Photo.

p. 7. [bottom] Rogers Fund, 1930 / The Metropolitan Museum of Art.

p. 8. Author's photos.

p. 11. Album / Alamy Stock Photo.

p. 16. Wellcome Collection. Public Domain.

p. 17. Roland and Sabrina Michaud / akg-images.

p. 20. Wellcome Collection. Public Domain.

p. 21. Wellcome Collection. Public Domain.

p. 23. Wellcome Collection. Public Domain.

p. 24. Courtesy of the New York Academy of Medicine Library.

p. 25. Courtesy of the New York Academy of Medicine Library.

p. 33. The Bookworm Collection / Alamy Stock Photo.

p. 36. By permission of the Master and Fellows of Gonville and Caius College.

p. 38. Photo: © Bodleian Libraries, University of Oxford.

p. 39. Science History Images / Alamy Stock Photo.

p. 42. World History Archive / Alamy Stock Photo.

p. 43. Wellcome Collection. Public Domain.

p. 46. Rosenwald Collection / National Gallery of Art.

p. 48. Carità by Giotto, fresco, Scrovegni (Arena) Chapel, Padua, Italy.

pp. 50–1. Vincenzo Fontana / Getty Images.

p. 52. Self-published work by G.dallorto / Wikimedia Commons.

p. 55. Bequest of Lore Heinemann, in memory of her husband, Dr. Rudolf J. Heinemann, 1996 / The Metropolitan Museum of Art, New York.

pp. 62–3. akg-images.

p. 64. Rogers Fund and The Cloisters Collection, by exchange, 1950 / The Metropolitan Museum of Art.

p. 76. Royal Collection Trust / © His Majesty King Charles III, 2024 / Bridgeman Images.

p. 79. Royal Collection Trust / © His Majesty King Charles III, 2024 / Bridgeman Images.

p. 82. Royal Collection Trust / © His Majesty King Charles III, 2024 / Bridgeman Images.

pp. 84–5. Royal Collection Trust / © His Majesty King Charles III, 2024 / Bridgeman Images.

p. 88. Royal Collection Trust / © His Majesty King Charles III, 2024 / Bridgeman Images.

p. 92. Royal Collection Trust / © His Majesty King Charles III, 2024 / Bridgeman Images.

p. 94–5. Royal Collection Trust / © His Majesty King Charles III, 2024 / Bridgeman Images.

pp. 96–7. Royal Collection Trust / © His Majesty King Charles III, 2024 / Bridgeman Images.

p. 99. Author's Work / Oxford University Press.

p. 106. Wellcome Collection (CC by 4.0).

pp. 110–111. The National Library of Medicine.

p. 112. Wellcome Collection. Attribution 4.0 International (CC BY 4.0).

p. 116. Harris Brisbane Dick Fund, 1938 / The Metropolitan Museum of Art, New York.

p. 117. Wellcome Collection. Public Domain.

p. 119. Granger / Bridgeman Images.

p. 126. The Stapleton Collection / Bridgeman Images.

p. 128. Wellcome Collection. Public Domain.

p. 142. Wellcome Collection. Public Domain.

Index